How to Pass

SECOND EDITION

HIGHER

Chemistry

John Anderson

HODDER
GIBSON
AN HACHETTE UK COMPANY

The Publishers would like to thank the following for permission to reproduce copyright material:

Photo credits

p. 33 © Kunal Mehta/Shutterstock.com; **p. 58** © Africa Studio – Fotolia.com; **p. 67** (top) © Marius Graf – Fotolia.com, (bottom) © baibaz – Fotolia.com; **p. 86** © Sheila Terry/Science Photo Library; **p. 150** © Rabbitmindphoto/Shutterstock.com

Acknowledgements

Questions, where marked throughout by an asterisk, are used by permission Copyright © Scottish Qualifications Authority. All questions, answers and worked examples without an asterisk do not emanate from SQA material.

Every effort has been made to trace all copyright holders, but if any have been inadvertently overlooked the Publishers will be pleased to make the necessary arrangements at the first opportunity.

Although every effort has been made to ensure that website addresses are correct at time of going to press, Hodder Gibson cannot be held responsible for the content of any website mentioned in this book. It is sometimes possible to find a relocated web page by typing in the address of the home page for a website in the URL window of your browser.

Hachette UK's policy is to use papers that are natural, renewable and recyclable products and made from wood grown in well-managed forests and other controlled sources. The logging and manufacturing processes are expected to conform to the environmental regulations of the country of origin.

Orders: please contact Bookpoint Ltd, 130 Park Drive, Milton Park, Abingdon, Oxon OX14 4SE. Telephone: (44) 01235 827827. Fax: (44) 01235 400454. Email: education@bookpoint.co.uk. Lines are open 9.00–5.00, Monday to Friday, with a 24-hour message answering service. Visit our website at www.hoddereducation.co.uk. If you have queries or questions that aren't about an order, you can contact us at hoddergibson@hodder.co.uk

© John Anderson 2019

First published in 2019 by
Hodder Gibson, an imprint of Hodder Education,
An Hachette UK Company
211 St Vincent Street
Glasgow G2 5QY

Impression number   5   4   3   2

Year   2023 2022 2021 2020

Cover photo © Olivierl/123RF
Illustrations by Aptara, Inc.
Typeset in 13/15 Cronos Pro (Light) by Aptara, Inc.
Printed in India
A catalogue record for this title is available from the British Library
ISBN: 978 1 5104 5231 2

MIX
Paper from
responsible sources
FSC™ C104740
www.fsc.org

SCOTLAND
EXCEL

We are an approved supplier on
the Scotland Excel framework.

Schools can find us on their
procurement system as:

**Hodder & Stoughton Limited t/a
Hodder Gibson.**

Contents

Introduction

Welcome to How To Pass Higher Chemistry, 2nd edition! This book follows on from the very successful 1st edition and has been updated using the 2018 SQA arrangements document for Higher Chemistry.

How to use this book

This book is designed to help you prepare for the SQA Higher Chemistry exam. It provides a complete summary of the Higher Chemistry course and offers the following features to aid understanding and improve recall of important points.

What you should know

This is provided at the start of each chapter, highlighting the SQA learning outcomes that you must know in order to demonstrate the required knowledge to meet the SQA assessment standards. Look at these before you read the chapter to make you think about what you already know and what you should find out more about. Once you have read the chapter and completed the questions, go back and check the *What you should know* points to check that you have mastered the required knowledge from this chapter.

Hints & tips

This feature gives you examiners' hints and tips on how to achieve top marks. It also highlights common mistakes to watch out for and avoid!

Key points

You will find the key points listed at the end of each chapter. These summarise the knowledge and skills required to achieve the SQA outcomes. You can use the key points to check your understanding of the chapter prior to testing your knowledge using the *Study Questions* at the end of the chapter.

Remember

This feature provides examiner's advice on specific points that you must memorise from the chapter. This could be a calculation skill (e.g. knowing how to calculate the relative rate of a reaction) or a specific piece of knowledge that you should be able to recall and explain (e.g. going across a period, the covalent radii decreases).

Examples

Each chapter contains worked examples to test your knowledge of the concepts in the chapter and to show you how best to answer Higher Chemistry questions. Try the question before you review the answer. If you cannot recall how to answer the question, try to search for the information in the chapter. Once you have exhausted your memory and have tried to solve the question, you can review the exemplar answer. This is an important point in learning: you must make every effort to recall before looking at an answer. The mental effort of searching for an answer helps to strengthen your memory and will boost your ability to learn.

Study questions

These are found at the end of the chapter and should be used to test your knowledge and understanding of the chemistry covered. The questions included have been carefully selected with many questions sourced from past SQA Higher exams. Like the worked examples, always exhaust all efforts to recall/search for an answer before checking using the answers at the back of the book.

Additional features of the Higher Chemistry exam

This appears in the appendix of the book and offers advice on how to tackle two important features of the Higher Chemistry exam: Numeracy and Open-ended questions. Both concepts are explored in detail and supported with worked examples and study questions. Take your time to review these features and, like previous chapters, test your knowledge and understanding using both the worked examples and study questions.

Glossary

A glossary of key words you should know is included at the end of the book. If you have reviewed each chapter and tested your knowledge using the advice given, you should be familiar with the key words and should know the definitions.

The Higher Chemistry course

The Higher Chemistry course comprises four areas of chemistry, which are presented as four sections in this *How to Pass* book. The four sections, and a brief summary of their contents, are:

Chemical changes and structure

This section explores the structure and bonding of elements in the Periodic Table and examines some of the properties of compounds that derive from their bonding.

Nature's chemistry

This section explores the chemistry of carbon compounds (a branch of chemistry known as organic chemistry) such as alcohols, esters and proteins. Their structures and reactions are explored with reference made to how chemical bonding affects the properties of the compounds.

The uses of the carbon compounds in areas such as cosmetics and the food industry are detailed, along with an explanation of how the compounds' properties are related to their structure.

Chemistry in society

This section examines the role of chemists in making new products for profit, covering concepts such as reaction rates, chemical energy and equilibria. The section concludes with a review of some of the techniques chemists use to analyse the compounds they have made.

Researching chemistry

This section is designed to summarise the practical chemistry techniques and apparatus you are likely to use in a chemistry laboratory at Higher level. A list of common practical techniques and equipment is detailed along with advice on best practice for carrying out such techniques.

Higher Chemistry assessment

There are three parts to the Higher Chemistry assessment:

1 An assignment
2 A multiple choice exam
3 An extended answer exam

The assignment is carried out as directed by your teacher/lecturer and will involve you researching an area of chemistry related to the Higher course. This will involve an experimental stage where you will conduct relevant practical work; a research stage where you will use books and the internet to find out about your chosen area of research; and a reporting stage where you will write a report summarising your findings. Your written report will be submitted to the SQA for marking and contributes 20% to the overall course mark.

The written exam, usually taken in May, comprises two question papers: a multiple-choice paper (**25 marks**) and an extended-answer paper (**95 marks**). You will be given **40 minutes** to answer the multiple-choice questions and **2 hours 20 minutes** to answer the extended-answer questions. Overall, the written exams contribute 80% to the overall course mark.

Advice for all Higher Chemistry students

The key to success in Higher Chemistry is to equip yourself with as much chemical knowledge as possible and to ensure that you can do all of the standard questions that you are going to be asked in the exam. In some ways, this is easy as this book presents you with all the chemical knowledge you need and includes the likely question types you will be asked. The tricky bit is learning and understanding this knowledge, and remembering how to do the standard questions. To do this, you need to get into the habit of reviewing the relevant chapters at the same time as you are studying the concepts in class. To review means to read and test that you understand and can recall. Using the worked examples and study questions will help you with this as they will force you to retrieve what you have read. The more you retrieve, the easier knowledge will come to you. In other words, you need to review and attempt the questions in each chapter several times.

It is also important to space your revision so that you leave some time between each review, as research shows that this is a more effective strategy than doing the same questions over and over again. So, an ideal strategy is to read a chapter, try the questions, check for understanding, and then move onto a different chapter. Go back to the original chapter a week or two later. Chances are, when you go back to review and try the questions again, you will have forgotten some of the information, which will force you to try to retrieve the information. This 'forgetting' and then 'retrieving' is an important part of revising and will help you remember the chemistry you should know.

Aside from the advice given in each chapter of this book, specific advice for tackling numeracy and open-ended question types is given in the appendix. Take time to review and apply this advice as it will serve you well when you tackle exam questions.

Finally, even if you decide not to pursue chemistry beyond the Higher course, I hope that studying chemistry has opened your eyes to the delights of the subject. It is a science that has many everyday applications from ensuring our indoor swimming pools are disinfected to the more complex field of designing new medicines to treat and cure disease.

Chemistry continues to have an exciting future in all areas of human endeavour from environmental conservation to the search for life-enhancing compounds. Many of you will, no doubt, continue your studies beyond Higher Chemistry and may, like many Higher Chemistry candidates, use your course knowledge to go on to advanced study of chemistry and its related disciplines. The knowledge and skills you acquire from this book will serve you well both now and in the future.

Best of luck!

Section 1 Chemical Changes and Structure

Chapter 1
Structure and bonding in the first 20 elements

What you should know

* ★ Elements are arranged in the Periodic Table in order of increasing atomic number.
* ★ The Periodic Table can be used to make predictions about the properties of elements. For example:
  * ★ Elements with the same number of outer electrons take part in similar chemical reactions.
  * ★ Elements can be classed as metals and non-metals and have corresponding properties.
* ★ The first 20 elements are categorised as:
  * ★ metallic (Li, Be, Na, Mg, Al, K, Ca)
  * ★ covalent molecular (H_2, N_2, O_2, F_2, Cl_2, P_4, S_8 and fullerenes)
  * ★ covalent network (B, C (diamond, graphite), Si)
  * ★ monatomic (noble gases).

All known elements are listed on the **Periodic Table**, which displays the elements in order of increasing **atomic number**. The Periodic Table can be used to make predictions about the properties of the elements. For example, all the group 1 elements (the alkali metals) react rapidly with water to form an alkaline solution and hydrogen gas. As you descend the group, the elements become increasingly reactive. This reactivity comes from the fact that all group 1 elements have 1 outer electron. As you will learn in the next chapter, as you descend the group, the outer electron becomes less attracted to the **nucleus** making it easier to remove, hence the increase in reactivity. Other trends in properties can be attributed to the outer electrons in elements such as the fact that some elements (towards the left-hand side of the table) are good conductors of electricity due to the fact that their outer electrons can move easily from atom to atom. At the opposite side of the table, the group 0 elements, the noble gases, are known to be unreactive because they have a full shell of outer electrons meaning that electrons are unlikely to be lost or gained from these atoms.

For the purposes of Higher Chemistry, you are expected to know the structure and bonding of the first 20 elements as illustrated in Figure 1.1. This chapter will explore some of the properties of these elements and will attempt to relate these properties to the structure and bonding of the elements.

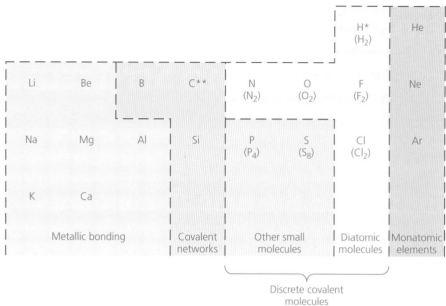

*Hydrogen is not a member of group VII
**Although unusually large, the fullerene
forms of carbon are discrete covalent molecules

Figure 1.1 Summary of the structure and bonding in the first 20 elements of the Periodic Table

Metallic bonding

Metallic bonding consists of positive metal **ions** surrounded by a pool of **delocalised electrons**. The attraction between the charged metal ions and the **electrons** is known as metallic bonding. As this is a relatively strong attraction, metals typically have high melting points reflecting the energy required to overcome the strong metallic bonds.

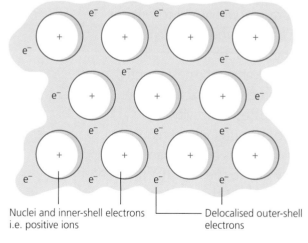

Nuclei and inner-shell electrons
i.e. positive ions

Delocalised outer-shell
electrons

Figure 1.2 Metallic bonding

A review of the SQA data booklet, page 6, allows you to compare the metallic bond strength of metal elements by comparing their melting points. For example, it can be seen that the melting point decreases as you descend the alkali metals: Li (181 °C), Na (98 °C) and K (63 °C) suggesting that the metallic bonds found in lithium are stronger than those found in sodium and potassium. In other words, the energy required to break the metallic bonds in lithium is greater than the energy required to break the metallic bonds in sodium and potassium. If you compare the group 2 metals to the group 1 metals on the same period, you will notice that the melting points of group 2 metals are much higher. For example, magnesium has a melting point of 650°C compared to 98°C for sodium.

The stronger metallic bonds found in magnesium are due to the fact that magnesium forms 2+ ions and contributes two electrons per atom to the delocalised pool of electrons. Both factors result in stronger metallic bonds in magnesium requiring a greater energy to overcome, hence the higher melting point than sodium (which forms 1+ ions and contributes one electron per atom to the delocalised pool of electrons).

A final feature of metallic bonding is the fact that the outer electrons are very loosely held. This means that the outer electrons are not bound to one ion: they can move around the metallic structure, i.e. they are delocalised. As the electrons are free to move, metals conduct electricity.

Covalent molecular elements

Hydrogen, nitrogen, oxygen and the halogens are all examples of diatomic molecules consisting of two atoms joined by **covalent bonding**. Phosphorus, sulfur and the fullerenes are also **covalent molecular** in structure, but consist of much larger molecules.

- Phosphorus consists of four phosphorus atoms joined together by covalent bonds as shown in Figure 1.3.

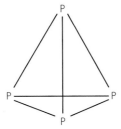

Figure 1.3 Phosphorus (P_4) molecule

- Sulfur can form molecules where eight sulfur atoms covalently bond to form 'puckered rings' as shown in Figure 1.4.

Figure 1.4 Bonding in sulfur (S_8)

- The fullerenes are a form of carbon consisting of five- and six-membered rings of carbon atoms covalently bonded together. One example of a fullerene consists of 60 carbon atoms, as shown in Figure 1.5.

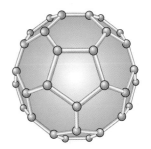

Figure 1.5 Fullerene structure (C_{60})

For all of the covalent molecular elements, the *intra*molecular forces, the bonds within the molecule, are covalent. The *inter*molecular forces, those between the molecules, are the very weak **London dispersion forces**. (London dispersion forces are discussed in greater detail in Chapter 3.) Most of these elements have relatively low melting and boiling points since only the weak London dispersion forces have to be broken to melt and boil them. The heavier molecules, such as sulfur, phosphorus and the fullerenes, have many more electrons than the lighter molecules. Consequently, there are stronger London dispersion forces between the molecules, resulting in the higher melting points of these elements.

Covalent network elements

Unlike covalent molecular substances, which consist of only a few atoms bonded together, **covalent network** structures consist of many thousands of atoms joined together by covalent bonds. These structures have very high melting points as strong covalent bonds must be broken for the solid to melt.

Carbon diamond

Diamond is one form of carbon which is an example of a covalent network structure. Each carbon atom is covalently bonded to four other carbon atoms in a tetrahedral arrangement as shown in Figure 1.6. The resultant structure is exceptionally hard and strong. As all four outer electrons are used to form covalent bonds to other carbon atoms, carbon diamond does not conduct electricity.

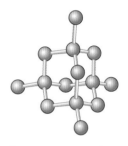

Figure 1.6 Diamond structure

Carbon graphite

In carbon graphite, each carbon atom forms three covalent bonds to neighbouring carbon atoms forming layers of hexagonal rings, as shown in Figure 1.7. The fourth outer electron becomes delocalised between the layers allowing carbon graphite to conduct electricity. The layers in graphite are held together by weak London dispersion forces. This allows the layers to move easily, making graphite an effective lubricant.

Remember

Covalent networks have very high melting points as strong covalent bonds must be broken when these structures are melted.

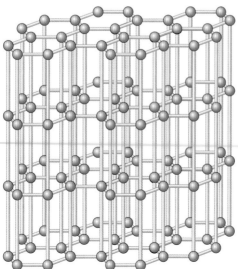

Figure 1.7 Graphite structure

Boron and silicon

Boron forms a covalent network structure based on B_{12} groups. Silicon forms a rigid covalent network with a similar tetrahedral structure to carbon diamond.

Monatomic elements

The noble gases are **monatomic** elements; they consist of single atoms which are not bonded to neighbouring atoms. When cooled, the atoms move closer together to form a liquid and then solid held together by weak London dispersion forces. As these forces are weak, the monatomic elements have very low melting points. As you descend the noble gases, the melting points increase as the London dispersion forces become stronger. This is due to the increased number of electrons.

Remember

London dispersion force strength increases as the number of electrons in an atom/molecule increase. For example, chlorine has a higher melting point than fluorine because the LDF between chlorine molecules is stronger than the LDF between fluorine molecules. This is because chlorine molecules have more electrons than fluorine molecules.

Hints & tips ★

If you are asked to explain why one element has a higher or lower melting point compared to another element, always refer to the type of bonds or forces that must be overcome to melt the element. Relate this to the amount of energy that has to be supplied. This is illustrated in the following worked examples.

Example

Explain why sodium is a solid at room temperature whereas chlorine is a gas at room temperature.

Solution

This is the same question as: why does sodium have a higher melting point than chlorine?

To melt sodium, strong metallic bonds must be broken. Sodium is a solid at room temperature as there is not enough energy (at room temperature) to break the strong metallic bonds.

Chlorine molecules are held together by weak London dispersion forces. At room temperature, enough energy is supplied to overcome the London dispersion forces allowing the chlorine molecules to separate as a gas.

Example

Explain why the boiling point of fluorine is higher than the boiling point of helium.

Solution

In both fluorine and helium, the force holding the fluorine molecules together and the force holding helium atoms together is London dispersion force.

The fact that the boiling point is higher for fluorine suggests that the LDF between fluorine molecules are stronger than the LDF between helium atoms. Thus, more energy is required to overcome the LDF between fluorine molecules than is required to overcome the LDF between helium atoms.

LDF strength increases as the number of electrons increases, therefore as there are more electrons in fluorine molecules than there are in helium atoms, the LDF between fluorine molecules is stronger than between helium atoms.

Key points

You should know how to describe the bonding and structures of the first 20 elements of the Periodic Table.

* Metallic: strong metallic bonds, high melting points, good conductors of electricity.
* Covalent molecular: strong covalent bonds between atoms; weak London dispersion forces between molecules; sulfur, phosphorus and the fullerenes are solids as they are heavier molecules with more electrons, therefore they have stronger London dispersion forces between molecules.
* Covalent network: strong covalent bonds between atoms; very high melting points as the strong covalent bonds must be broken; examples include carbon (diamond and graphite), silicon and boron.
* Monatomic: the noble gases are gaseous as they have weak London dispersion forces between atoms; melting points increase as you descend the group as the atoms have more electrons and therefore stronger London dispersion forces.
* You should be familiar with using the data booklet to look up information about the elements such as the electron arrangements, atomic numbers, melting and boiling points.
* Trends in reactivity of elements can be related to the number of outer electrons.
* Trends in melting and boiling points can be related to the structure and bonding of the elements.

Study questions ?

*1 Complete the following table by adding in the letter for the type of element, shown in Figure 1.8, which matches the description in Table 1.1.

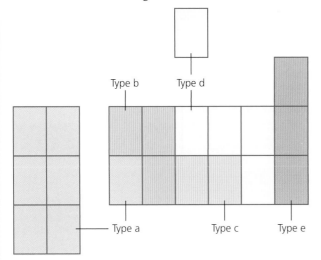

Figure 1.8

Table 1.1

Type		Description
1		Monatomic gases
2		Molecular solids
3		Covalent network solids
4		Metallic solids
5		Covalent molecular gases

2 Which of the following statements can be applied to both silicon and sulfur?
   A the high melting point is a result of strong covalent bonding
   B molecules are held together by LDF
   C atoms are held together by strong covalent bonds
   D the high melting point is a result of strong LDF between atoms

3 When liquid oxygen evaporates
   A covalent bonds are formed
   B covalent bonds are broken
   C London dispersion forces are broken
   D London dispersion forces are formed.

4 Which element would require covalent bonds to be broken when it is melted?
   A helium
   B nitrogen
   C boron
   D sodium

5 Which of the following elements will not conduct electricity?
   A potassium
   B carbon graphite
   C carbon diamond
   D sodium

6 **a)** Copy and complete Table 1.2 by adding the name of an element from elements 1 to 20 of the Periodic Table for each of the types of bonding and structure described.

Table 1.2

Bonding and structure at room temperature and pressure	Name of element
Metallic solid	Sodium
Monatomic gas	
Covalent network solid	
Discrete covalent molecular gas	
Discrete covalent molecular solid	

b) Why do metallic solids such as sodium conduct electricity?

7 Describe the structure of carbon diamond and carbon graphite. Explain why carbon diamond does not conduct electricity yet carbon graphite does conduct electricity.

Covalent radius

The **covalent radius** is a measure of the size of an atom. It is half the distance between the nuclei of two covalently bonded atoms of an element, as shown in Figure 2.1.

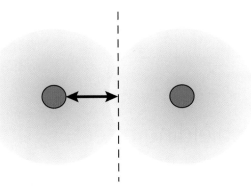

Figure 2.1 Half the distance between two nuclei is the covalent radius.

There are two general trends that are considered in Higher Chemistry:

Remember

1 Going across a **period**, covalent radius decreases.
2 Going down a **group**, covalent radius increases.

Covalent radius data can be obtained from the SQA data booklet, page 7.

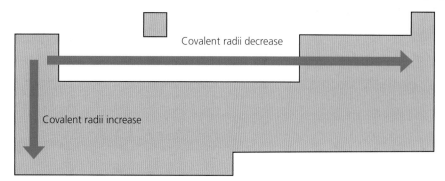

Covalent radii decrease

Covalent radii increase

Figure 2.2 Trends in covalent radius

The two factors that help explain these trends are the nuclear charge of the atom and the number of filled electron shells.

Going across a period: the effect of nuclear charge

Going across a period, the nuclear charge increases but the number of filled electron shells remains the same. This is shown in Table 2.1.

Table 2.1 The covalent radius decreases going across a period. Note that the unit of measurement here is picometres (pm), which is 1×10^{-12} m.

Element	Li	Be	B	C	N	O	F
Atomic number	3	4	5	6	7	8	9
Nuclear charge	3+	4+	5+	6+	7+	8+	9+
Electron arrangement	2,1	2,2	2,3	2,4	2,5	2,6	2,7
Covalent radius/pm	134	129	90	77	75	73	71

Remember

An increase in the nuclear charge results in electrons being more strongly attracted to the nucleus which means that the covalent radius decreases.

Going down a group: the shielding effect

Going down a group, the number of filled electron shells increases. This is shown in Table 2.2.

Table 2.2 The covalent radius increases going down a group.

Element	Li	Na	K	Rb	Cs
Atomic number	3	11	19	37	55
Nuclear charge	3+	11+	19+	37+	55+
Electron arrangement	2,1	2,8,1	2,8,8,1	2,8,18,8,1	2,8,18,18,8,1
Covalent radius/pm	134	154	196	216	235

Remember

As you go down a group, an extra shell of electrons is added. This explains why the covalent radius increases as you go down a group.

As you go down a group, the nuclear charge increases too. You would expect this to pull electrons in closer to the nucleus. However, the effect of increasing nuclear charge is outweighed by the much greater effect of adding extra shells of electrons. Each extra layer of electrons 'shields' the outer electrons from the positive nucleus so that the outer electrons are less strongly attracted to the nucleus. This shielding effect is also known as **screening**.

Example

Explain the trend in atomic size as you go across a period in the Periodic Table.

Solution

The atomic size, as measured by the covalent radius, decreases as you go across a period. This is due to the nuclear charge increasing causing electrons to be pulled closer to the nucleus.

Example

Explain why the covalent radius of a potassium atom is larger than the ionic radius of a potassium ion.

Solution

The electron arrangement for a K atom is 2,8,8,1.

The electron arrangement for a K ion is 2,8,8.

This shows that the K atom is larger as it has four electron shells, whereas the K ion has only three electron shells.

Hints & tips

You may be asked to compare trends for ions instead of atoms. To answer such questions, remember that negative ions have gained electrons, whereas a positive ion has lost electrons. Use this fact to help you write the electron arrangement for the ion. Look at the following worked example for guidance.

Electronegativity

Electronegativity is a measure of attraction for electrons in a bond. The higher the electronegativity value of an element, the stronger its attraction for electrons. For example, in a molecule of hydrogen chloride, the Cl attracts the shared electrons much more strongly than the H. Cl has an electronegativity value of 3.0 whereas H has a value of 2.2.

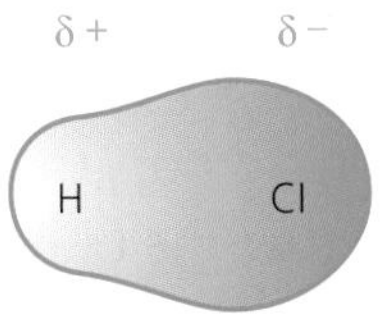

The trends in electronegativity are:

$\delta+$ $\delta-$

Figure 2.3 The Cl atom attracts the bonded electrons more strongly than the H atom in a molecule of hydrogen chloride.

Remember

1 *Going down a group, electronegativity decreases.*
2 *Going across a period, electronegativity increases.*

Electronegativity data can be obtained from the SQA data booklet, page 11.

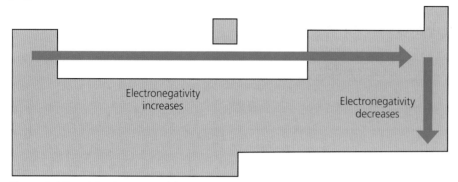

Figure 2.4 Trends in electronegativity

As with covalent radius, the two factors that influence electronegativity are the nuclear charge of an atom and the number of filled electron shells.

Table 2.3 Electronegativity increases across a period and decreases down a group.

H 2.2							
Li 1.0	Be 1.5	B 2.0	C 2.5	N 3.0	O 3.5	F 4.0	decrease down group
Na 0.9	Mg 1.2	Al 1.5	Si 1.9	P 2.2	S 2.5	Cl 3.0	
K 0.8	Ca 1.0	Ga 1.8	Ge 2.0	As 2.2	Se 2.4	Br 2.8	
Rb 0.8	Sr 1.0	In 1.7	Sn 1.8	Sb 2.1	Te 2.1	I 2.6	
Cs 0.8	Ba 0.9						

increase across period

Going across a period, the nuclear charge increases. This increase in nuclear charge causes the atom to attract bonded electrons more strongly. Consequently, electronegativity increases across a period.

Going down a group, the number of filled electron shells increases. This means that the outer electrons are further from the nucleus and are less strongly attracted to the nucleus. In addition, the extra shells of electrons screen the outer electrons from the nuclear charge which means that they are less strongly attracted to the nucleus. Consequently, electronegativity decreases going down a group.

Example

Explain the trend in electronegativity as you descend the halogens, group 7.

Solution

As you go down group 7, the electronegativity decreases.

As you go down the group, an extra layer of electrons is added. For example, F is 2,7; Cl is 2,8,7; Br is 2,8,18,7, etc. This results in the outer electrons (the electrons that will form covalent bonds) being increasingly distant from the nucleus. In addition, the extra layers of filled electron shells will shield the outer electrons from the nuclear charge. Overall, this results in the outer electrons being less strongly attracted to the nucleus. Hence, electronegativity, which is a measure of attraction for bonded electrons, decreases as you go down group 7.

Ionisation energy

The **ionisation energy** is defined as 'the energy required to remove one **mole** of electrons from one mole of gaseous atoms'.

$$Na(g) \rightarrow Na^+(g) + e^- \qquad \Delta H = 496 \, kJ \, mol^{-1}$$

For sodium, 496 kJ of energy is required to remove the first electron from one mole of sodium atoms in the gaseous state. This is known as the first ionisation energy of sodium since it is a measure of the energy required to remove the first, or outermost, electron from sodium.

The second ionisation energy is the energy required to remove a second electron from sodium after the first electron has been removed.

$$Na^+(g) \rightarrow Na^{2+}(g) + e^- \qquad \Delta H = 4562 \, kJ \, mol^{-1}$$

The trends in first ionisation energy are:

Remember

1 Going down a group, the ionisation energy decreases.
2 Going across a period, the ionisation energy increases.

Ionisation energy data can be obtained from the SQA data booklet, page 7.

As with covalent radius and electronegativity, the factors that must be considered to help us explain the trends in ionisation energy are the nuclear charge of the atom and the number of filled electron shells.

Going across a period, the nuclear charge is increasing. The outermost electrons are therefore more strongly held and so the energy required to remove them, the ionisation energy, increases along each period.

Going down a group, an electron is being removed from the layer of electrons which is furthest from the nucleus. This layer is increasingly distant from the nuclear attraction and hence, although the nuclear charge is also increasing, less energy is required to remove an electron. An additional factor is the screening effect of electrons in inner shells. These inner electrons reduce the attraction of the nucleus for outermost electrons, hence reducing the ionisation energy.

Example

Explain why potassium has a lower first ionisation energy than lithium.

Solution

K: nuclear charge 19+; electron arrangement 2,8,8,1

Li: nuclear charge 3+; electron arrangement 2,1

Potassium has two extra filled layers of electrons compared to lithium. Consequently, the outer electron in K is further from the nucleus than the outer electron in Li. In addition, the extra layers of electrons in K screen the outer electron from the nucleus. Therefore, the outer electron in K is not as strongly attracted (it is weakly held) as the outer electron in Li. Overall, it will take less energy to remove the outer electron from K than will be required to remove the outer electron from Li.

Example

Explain why the second ionisation energy of sodium is much higher than the first ionisation energy of sodium.

Solution

Na: electron arrangement 2,8,1

Removing the second electron from sodium involves breaking into the second shell of electrons which is much closer to the nucleus. Consequently, more energy is required as the electron in the second shell is more strongly attracted to the nucleus.

Key points

* Trends for covalent radius, ionisation energy and electronegativity can be explained by comparing the nuclear charge or number of electron shells.
* Going down a group, the number of electron shells increases. This causes the size of the atom to increase. Since the outer electrons are now further from the nucleus, they are less strongly attracted to the nucleus. In addition, the extra layers of electrons added shield outer electrons from the nuclear charge. Both effects result in ionisation energy and electronegativity to decrease as you go down a group. ⇨

 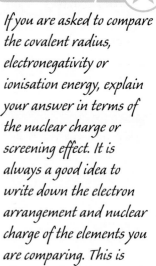

* Going across a period, the nuclear charge increases. This increases attraction for electrons resulting in ionisation energy and electronegativity increasing. Pulling electrons closer to the nucleus also results in atoms becoming smaller.

Study Questions

1 Which of the following elements will form a non-polar covalent bond when bonded to hydrogen?
   A phosphorus
   B fluorine
   C sodium
   D oxygen

2 A lithium atom is smaller than a sodium atom because lithium has
   A a lower nuclear charge
   B fewer layers of electrons
   C a higher electronegativity
   D a higher first ionisation energy.

3 Which of the following shows the correct equation for the second ionisation energy of potassium?
   A $K(g) \rightarrow K^{2+}(g) + 2e^-$
   B $K^+(g) \rightarrow K^{2+}(g) + 2e^-$
   C $K^+(g) \rightarrow K^{2+}(g) + e^-$
   D $K^{2+}(g) \rightarrow K^{3+}(g) + e^-$

4 Which statement correctly describes the reason for iodine having a larger covalent radius than fluorine?
   A Iodine has a higher nuclear charge.
   B Iodine has more layers of electrons.
   C Iodine has a higher first ionisation energy.
   D Iodine is more reactive than fluorine.

5 a) Why does the ionisation energy increase across a period?
   b) Write an equation corresponding to the first ionisation energy of sodium.
   c) In which group would you find the elements with the highest electronegativity values?

6 a) Aluminium and phosphorus are close to one another in the Periodic Table but the P^{3-} ion is much larger than the Al^{3+} ion. Give the reason for this difference.
   b) The P^{3-} ion and the Ca^{2+} ion have the same electron arrangement but the Ca^{2+} ion is smaller than the P^{3-} ion. Give the reason for this difference.

Chapter 3
Bonding

Structure and bonding

Pure covalent bonding

Diatomic elements such as hydrogen (H_2) exist as two atoms covalently bonded together. In other words, the two atoms *share* electrons. In the case of diatomic elements, both atoms have an equal 'pull' on the shared electrons (they have the same electronegativity) so we say that the bond is a *pure covalent bond* or a **non-polar covalent bond**.

Pure covalent bonding occurs in compounds where both atoms have the same electronegativity. For example, in NCl_3, both N and Cl have an electronegativity value of 3.0 so the bond between the N and Cl is a non-polar covalent bond.

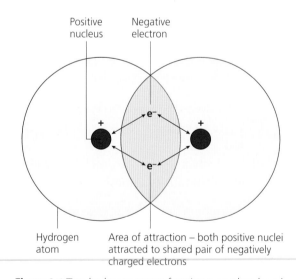

Figure 3.1 Two hydrogen atoms forming a covalent bond. This is a non-polar covalent bond since both H atoms have an equal attraction for the electrons in the bond.

Polar covalent bonding

In most compounds, the two atoms forming the covalent bond have different electronegativity values. In this case, the atom with the highest electronegativity attracts electrons more strongly than the other atom. This results in the atom with the higher electronegativity having a slight negative charge ($\delta-$) and the other atom having a slight positive charge ($\delta+$). This type of covalent bond is known as a **polar covalent bond**. In water, for example, the oxygen atom has a higher electronegativity than the hydrogen atom. In hydrogen chloride, the chlorine has a higher electronegativity than the hydrogen. This is illustrated in Figure 3.2.

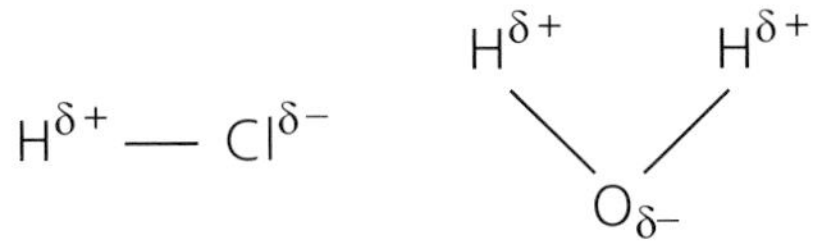

$$H^{\delta+} \text{---} Cl^{\delta-}$$

$$H^{\delta+} \qquad H^{\delta+}$$
$$O_{\delta-}$$

Figure 3.2 Hydrogen chloride and water have polar covalent bonds. This is caused by the atoms sharing the electrons having different electronegativity values.

Ionic bonding

Atoms with a large difference in electronegativity will sometimes form an **ionic bond**. An ionic bond does not involve sharing electrons; it occurs where electrons are transferred from one atom to another causing one atom to lose electrons (and become positively charged) and the other atom to gain electrons (and become negatively charged). The attraction between the positive ions of one element and the negative ions of the other element is known as an ionic bond. Typically, when a metal bonds with a non-metal, an ionic bond forms as the metal has a low electronegativity value and the non-metal has a much higher electronegativity value. This is a general rule and it should be remembered that some metal compounds will be covalent.

Ionic structure and ionic formula

Ionic compounds form a structure known as an ionic lattice, as shown in Figure 3.3.

This structure is composed of many millions of oppositely charged ions. The formula for an ionic compound tells us the ratio of ions in the lattice. For example, the sodium chloride formula Na^+Cl^- tells us that for every sodium ion there is one chloride ion. The formula for magnesium chloride is $Mg^{2+}(Cl^-)_2$. This tells us that there are two chloride ions for every one magnesium ion. Thus, if you knew that there were 2 million magnesium ions in the lattice, you would be able to predict that there should be 4 million chloride ions present too.

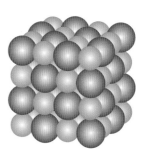

Figure 3.3 An ionic lattice structure

The bonding continuum

Some ionic compounds have stronger ionic properties than others and some covalent compounds have stronger covalent properties than others. When comparing compounds, the one with the greatest difference in electronegativity would usually be the 'most' ionic. The concept of a **bonding continuum** can be used to help us appreciate the differences in bonding, where pure ionic bonding and pure covalent bonding are at opposite ends and polar covalent bonding is in the middle. This idea is illustrated in Figure 3.4 (on page 18).

Compound	LiF	BeF_2	NF_3	OF_2	(F_2)
Difference in electronegativity	3.0	2.5	1.0	0.5	0.0

Ionic   Polar covalent   Covalent

Figure 3.4 The bonding continuum

Ionic or covalent bonding?

Electronegativity is one tool for deciding the type of bonding in a compound, but it is important to look also at the properties of the compound. There are three main characteristics of compounds that can be used to help decide whether the compound being studied is ionic or covalent:

1 Ionic compounds will conduct electricity when molten or when they are dissolved in water; covalent compounds will not conduct.
2 Ionic compounds tend to have high melting points as a lot of energy is required to break the strong ionic bonds that exist in the ionic **lattice** formed by such compounds in the solid state.
3 Ionic compounds are usually soluble in water.

The melting point of covalent compounds varies enormously as covalent compounds can exist in huge network structures with very high melting points (such as SiO_2, mp 1610 °C) or they can exist as small molecules with much lower melting points (such as CH_4, mp −182.5 °C).

Tin (IV) iodide is an example of a low melting point solid (144 °C) that does not conduct electricity when molten. Given this information we can conclude that tin (IV) iodide is a covalent compound. The fact that tin has an electronegativity value of 1.8 and iodine has an electronegativity value of 2.6 allows us to conclude that tin (IV) iodide contains polar covalent bonds.

Remember

Not all metal compounds are ionic. You have to look at the properties of the compound to decide if it contains ionic bonds.

Example

A student stated that the ionic formula for $Ca^{2+}(F^-)_2$ told her that each molecule contained 1 calcium atom and 2 fluorine atoms. Comment on the accuracy of the student's statement.

Solution

There are two inaccurate statements:

1 The use of the term 'molecule' since ionic compounds do not exist as small molecules. They exist in huge structures known as the ionic lattice.
2 The student suggests that the formula tells her that each molecule contains 1 calcium atom and 2 fluorine atoms. The formula actually tells us the ratio of ions (not atoms). Thus, we can say that there are 2 fluoride ions to every 1 calcium ion.

Intermolecular forces and properties of compounds

What you should know

Intermolecular forces

★ All molecular elements and compounds and monatomic elements condense and freeze at sufficiently low temperatures. For this to occur, some attractive forces must exist between the molecules or discrete atoms.

★ Intermolecular forces acting between molecules are known as van der Waals' forces. There are several different types of these, such as London dispersion forces and permanent dipole–permanent dipole interactions that include hydrogen bonding.

★ London dispersion forces are forces of attraction that can operate between all atoms and molecules. These forces are much weaker than all other types of bonding. They are formed as a result of electrostatic attraction between temporary dipoles and induced dipoles caused by movement of electrons in atoms and molecules.

★ The strength of London dispersion forces is related to the number of electrons within an atom or molecule.

★ A molecule is described as polar if it has a permanent dipole.

★ The spatial arrangement of polar covalent bonds can result in a molecule being polar.

★ Permanent dipole–permanent dipole interactions are additional electrostatic forces of attraction between polar molecules.

★ Permanent dipole–permanent dipole interactions are stronger than London dispersion forces for molecules with similar numbers of electrons.

★ Bonds consisting of a hydrogen atom bonded to an atom of a strongly electronegative element, such as fluorine, oxygen or nitrogen, are highly polar. Hydrogen bonds are electrostatic forces of attraction between molecules that contain these highly polar bonds. A hydrogen bond is stronger than other forms of permanent dipole–permanent dipole interaction but weaker than a covalent bond.

★ Melting points, boiling points, viscosity and solubility can all be rationalised in terms of the nature and strength of the intermolecular forces that exist between molecules. By considering the polarity and number of electrons present in molecules, it is possible to make predictions of the strength of the intermolecular forces.

★ The melting and boiling points of polar substances are higher than the melting and boiling points of non-polar substances with similar numbers of electrons.

★ The anomalous boiling points of ammonia, water and hydrogen fluoride are a result of hydrogen bonding.

★ Hydrogen bonding between molecules in ice results in an expanded structure that causes the density of ice to be less than that of water at low temperatures.

★ Ionic compounds and polar molecular compounds tend to be soluble in polar solvents such as water, and insoluble in non-polar solvents. Non-polar molecular substances tend to be soluble in non-polar solvents and insoluble in polar solvents.

★ To predict the solubility of a compound, key features to be considered are the
  ★ presence in molecules of O–H or N–H bonds, which implies hydrogen bonding
  ★ spatial arrangement of polar covalent bonds, which could result in a molecule possessing a permanent dipole.

Ionic compounds are held together in the solid state by ionic bonds. Covalent network compounds are held together in the solid state by covalent bonds. Covalent molecules, such as water or carbon dioxide, are held together in the solid state by intermolecular forces of attraction known as **van der Waals' forces**. When we heat a solid compound such as water (ice), energy is required to break the van der Waals' forces which hold the molecules together.

There are three main types of van der Waals' force:

1 London dispersion forces
2 permanent dipole–permanent dipole interactions
3 hydrogen bonding.

Figure 3.5 summarises the properties of the three types of forces.

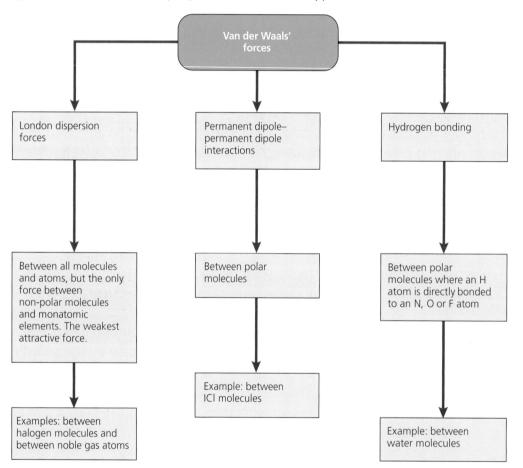

Figure 3.5 A summary of van der Waals' forces

London dispersion forces (LDF)

London dispersion forces are the weakest forces of attraction which can operate between atoms and molecules. They are caused by the uneven distribution of moving electrons. Figure 3.6 illustrates how this occurs in a monatomic element. The same principle can be applied to molecules.

Figure 3.6 shows the formation of **temporary dipoles** in which the side of the atom which has an excess of electrons becomes $\delta-$, causing the other side of the atom to become $\delta+$. The electrons in

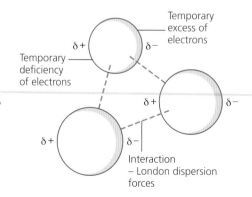

Figure 3.6 London dispersion forces

a neighbouring atom will shift away from an approaching δ– causing a δ+ to appear. This is known as an *induced dipole* since one atom has caused this to happen to its neighbour. This sets up the attraction we call London dispersion forces.

London dispersion forces are the main force between non-polar molecules.

Permanent dipole–permanent dipole interactions (pdp–pdp)

Permanent dipole–permanent dipole interactions occur between polar molecules and are much stronger than London dispersion forces. This can be illustrated by examining the polar molecule known as propanone, shown in Figure 3.7.

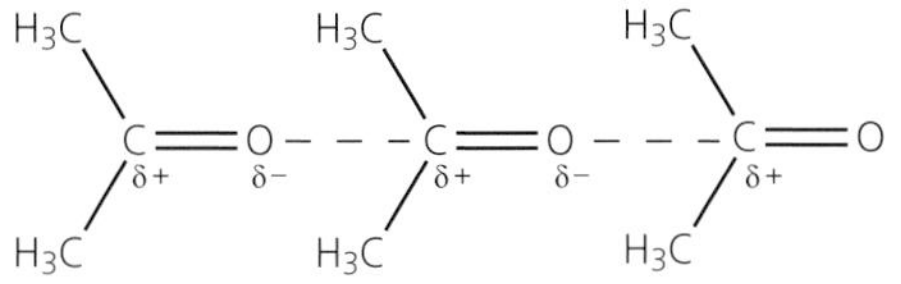

Figure 3.7 Propanone molecules bond to each other by pdp–pdp interactions.

The permanent dipole in one propanone molecule is attracted to the permanent dipole in a neighbouring propanone molecule. We know that this is stronger than London dispersion forces as the melting and boiling points of compounds that contain pdp–pdp interactions are much higher than those that contain London dispersion forces. For example, propanone can be compared to butane as both molecules have a similar number of electrons. Therefore, any difference in melting or boiling point must be due to an intermolecular force other than London dispersion forces.

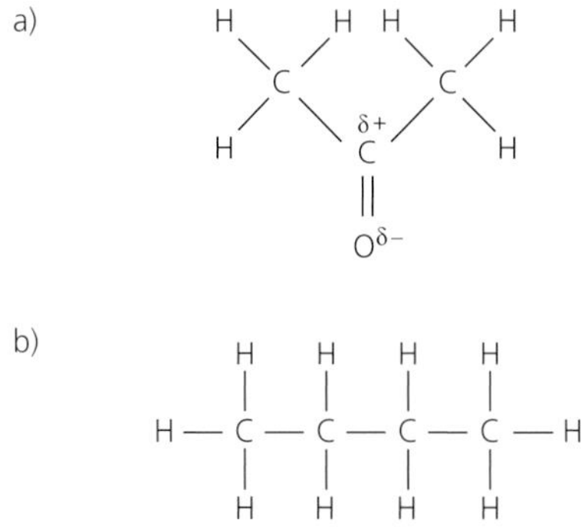

Figure 3.8 a) Propanone: bp 56 °C; b) butane: bp 0 °C

Propanone has a much higher boiling point than butane. This tells us that it takes much more energy to break apart the attractions between propanone molecules than it does to break apart the attractions between butane molecules. Since propanone has pdp–pdp interactions and butane has LDF between molecules, the higher boiling point of propanone tells us that pdp–pdp interactions are much stronger than LDF.

Polar or non-polar molecules?

Most compounds that contain polar covalent bonds are, overall, polar molecules. In other words, the molecules that make up the compounds have a permanent dipole where one side of the molecule is δ+ and the other side of the molecule is δ–. Ammonia is a good example and is illustrated in Figure 3.9.

Hints & tips

London dispersion forces are strongest between atoms (or molecules) with the most electrons. This is illustrated when comparing the noble gases, which are held together in the solid state by London dispersion forces. As you descend the noble gases, the melting point increases as it takes more energy to overcome the London dispersion forces between the atoms. The London dispersion forces are stronger because the number of electrons increases as you descend the group.

Remember

LDF are caused by temporary dipoles whereas pdp–pdp are caused by permanent dipoles.

LDF is the weakest type of intermolecular force.

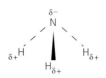

Figure 3.9 Ammonia is an example of a polar molecule.

Other molecules have a symmetrical arrangement of polar bonds causing the polarity to cancel. Carbon dioxide, **hydrocarbons** such as the **alkanes**, and tetrachloromethane are examples of non-polar molecules that have polar bonds which cancel.

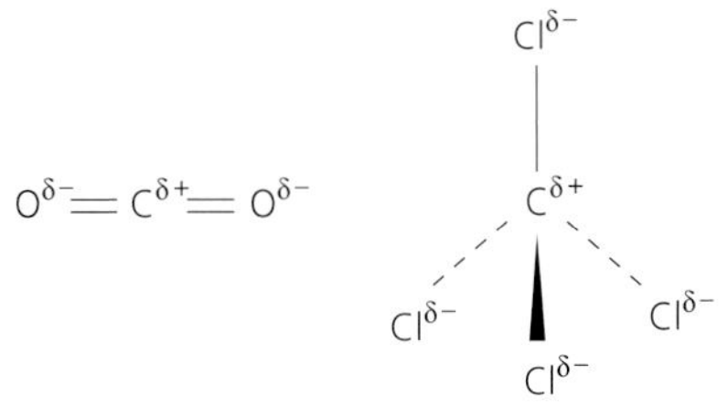

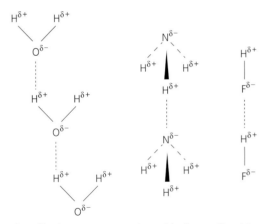

Figure 3.10 These molecules – carbon dioxide and tetrachloromethane – contain polar bonds but are, overall, non-polar molecules.

Hydrogen bonding

Hydrogen bonding is the strongest of the three intermolecular attractions. It occurs between molecules where there is an atom of H joined to an atom of N, O or F. The dashed lines in Figure 3.11 represent the hydrogen bonds between the molecules. Because these attractions are much stronger, hydrogen-bonded compounds have much higher melting and boiling points than would be expected.

Figure 3.11 Hydrogen bonding in water, ammonia and hydrogen fluoride

Hints & tips

Non-polar molecules will be held together by London dispersion forces whereas polar compounds will be held together by pdp–pdp interactions or hydrogen bonds.

Example

Which of the following molecules is likely to be non-polar?

A water
B propan-1-ol
C tetrachloromethane
D hydrogen chloride

Solution

The answer is **C** since the polarities cancel in tetrachloromethane as shown in Figure 3.10.

Relating properties to intermolecular forces

Melting and boiling points

Differences in intermolecular forces of attraction give rise to compounds having different melting and boiling points. This is illustrated in the examples which follow.

Octane (C_8H_{18}) and methane (CH_4) are both non-polar covalent molecules. The main force of attraction between non-polar molecules is London dispersion forces. Octane has a much higher boiling point because it has more electrons and therefore the strength of LDF between octane molecules is greater than the strength of LDF between methane molecules. Consequently, more energy is required to break the LDF attraction between octane molecules and hence octane has a higher boiling point.

Hydrogen sulfide (H_2S) and water (H_2O) are both examples of polar molecules. The boiling point of water is much higher than the boiling point of hydrogen sulfide because water molecules are held together by hydrogen bonds whereas hydrogen sulfide molecules are held together by pdp–pdp interactions. Hydrogen bonding is stronger than pdp–pdp interactions, therefore it takes more energy to break apart the hydrogen bonds between water molecules than it does to break apart the pdp–pdp interactions between hydrogen sulfide molecules. Consequently, water has a higher boiling point.

Solubility

In general, polar solvents such as water can dissolve polar and ionic substances. (A polar solvent is one in which the ends of the molecules have slight positive ($\delta+$) and negative ($\delta-$) charges.) Polar and ionic substances will not dissolve in non-polar substances. For example, salt (NaCl) will dissolve easily in water but will not dissolve in the non-polar solvent heptane.

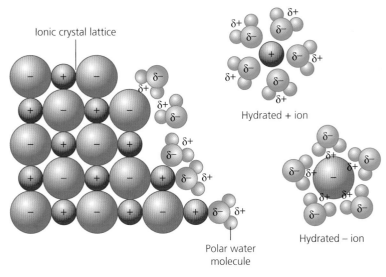

Figure 3.12 Water can dissolve an ionic compound as it is able to attract the ions as shown.

Non-polar compounds can dissolve in non-polar solvents but cannot dissolve in polar solvents. For example, wax (a non-polar hydrocarbon) can dissolve in hexane (a non-polar solvent) but will not dissolve in water (a polar solvent).

Example

Which two of the following compounds are likely to be soluble in water?

A ethanol   **C** PH_3

B hexane   **D** LiCl

Solution

Compound A is a polar molecule so is likely to dissolve in water.

Compound B is a non-polar molecule so will not dissolve in water.

Compound C is a non-polar molecule so will not dissolve in water.

Compound D is an ionic compound so is likely to dissolve in water.

Example

Table 3.1

Compound	Formula	Boiling point, °C
Ammonia	NH_3	−33
Phosphine	PH_3	−88

Despite having more electrons per molecule, phosphine has a lower boiling point than ammonia. Explain this difference.

Solution

Ammonia contains an N–H bond, which is highly polar. Overall, ammonia molecules are polar and can form hydrogen bonds to other ammonia molecules (as shown in Figure 3.11). Phosphorus and hydrogen have the same electronegativity value (2.2). Therefore, molecules of phosphine are non-polar. This means that the intermolecular force between phosphine molecules is the London dispersion force. Since hydrogen bonding is stronger than LDF, it will take much more energy to break the H bonds between ammonia molecules than will be required to break the LDF between phosphine molecules. Hence, ammonia has a higher boiling point.

Viscosity

Viscous liquids have strong intermolecular forces between molecules. For example, when **glycerol** and ethanol are compared (Figures 3.13 and 3.14), glycerol is found to be a much more viscous liquid.

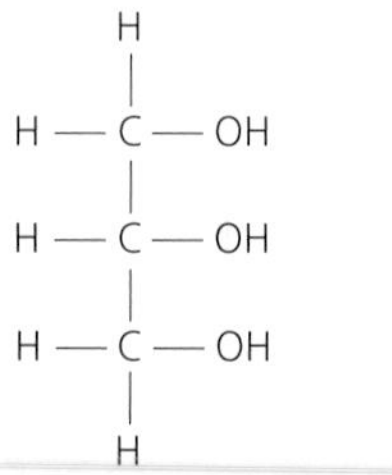

Ethanol has one hydroxyl (–OH) group whereas glycerol has three –OH groups. Consequently, hydrogen bonding between glycerol molecules is much stronger than the hydrogen bonding that can occur between ethanol molecules. Overall, increasing the number of **hydroxyl groups** in a molecule increases the **viscosity**.

Propane-1,2,3-triol or glycerol

Figure 3.13 Glycerol

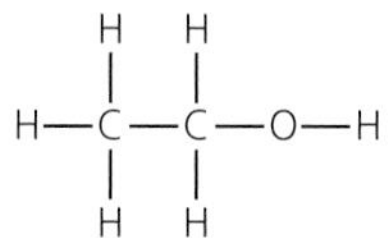

Figure 3.14 Ethanol

The density of ice

When most liquids become solids, the density of the solid is greater than that of the liquid since the particles in the solid state are packed much closer together compared to the particles in the liquid state. Water is unusual in that at its freezing point (the point at which it turns to ice), it is less dense than liquid water. This arises because the most efficient hydrogen bonding in ice comes from the water molecules adopting an arrangement known as an 'open-lattice' structure which results in lots of spaces between water molecules. The fact that ice is less dense than liquid water allows the solid ice to float on top of water.

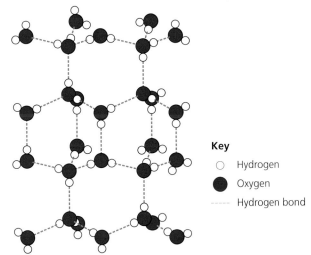

Key

○ Hydrogen

● Oxygen

----- Hydrogen bond

Figure 3.15 The open-lattice arrangement of water molecules in ice

Key points !

* Ionic compounds are formed when two atoms bond where there is a large difference in electronegativity. They can be identified by their properties which include high melting points, conductors of electricity when molten, and soluble in polar solvents.
* Covalent compounds are classed as polar or non-polar.
* Polar covalent bonds are formed when two atoms with different electronegativity values form a covalent bond. The atom with the higher electronegativity value is assigned the symbol δ– and the atom with the lower value is assigned the symbol δ+.
* The bonding continuum is used to describe the variation in bonding from ionic at one end to covalent at the other end. Polar covalent is in the middle of the continuum.
* The attractions between covalent molecules are known as van der Waals' forces.
* Three types of van der Waals' forces are studied in this chapter. They are, from weakest to strongest: London dispersion forces (LDF), permanent dipole–permanent dipole interactions (pdp–pdp), and hydrogen bonds.
* Non-polar covalent bonds are formed between two atoms with the same electronegativity value. The atoms have an equal attraction for electrons in the covalent bond.
* Compounds with polar bonds will form polar molecules provided the molecule is not symmetrical.
* Compounds with non-polar bonds will form non-polar molecules.
* Symmetrical compounds with polar bonds (such as CO_2) will form non-polar molecules.
* Non-polar molecules are attracted to other non-polar molecules by LDF.
* Polar molecules are attracted to other polar molecules by pdp–pdp interactions or hydrogen bonding.
* Permanent dipole–permanent dipole interactions occur between molecules where the molecule has a permanent dipole, for example, between ICl molecules. ⇨

* Hydrogen bonding occurs between molecules that contain an atom of H bonded to an atom of N, O or F, for example, between alcohol molecules.
* Molecules with stronger intermolecular forces will have higher melting and boiling points and a higher viscosity.
* Solubility can be predicted using the rule 'like dissolves like'. In other words, polar molecules and ionic substances will dissolve in polar solvents; non-polar compounds will dissolve in non-polar solvents.

Study questions

1 Which of the following compounds contains a polar covalent bond?

  A NaCl   **C** CO_2

  B CS_2 **D** NCl_3

2 Which of the following compounds is a non-polar covalent compound that contains a polar covalent bond?

  A $MgCl_2$ **C** HCl

  B PH_3 **D** CCl_4

3 Which of the following bonding types is never found in elements?

  A covalent network   **C** non-polar covalent

  B covalent molecular   **D** polar covalent

4 Which of the following solvents is least likely to dissolve a sample of sodium chloride?

  A water   **C** ethanol

  B cyclohexane   **D** ammonia (NH_3)

5 A non-polar covalent bond will be formed between nitrogen and

  A chlorine   **C** fluorine

  B hydrogen   **D** oxygen.

6 Which of the following properties would never be found in an ionic compound?

  A It is soluble in water.   **C** It conducts as a solid.

  B It has a high melting point.   **D** It is highly coloured.

7 Figure 3.16 shows the trend in boiling points for four alkanes. Explain the trend.

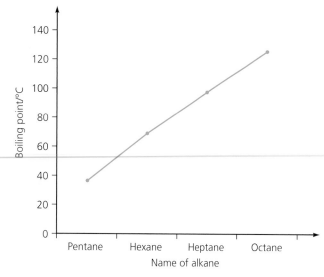

Figure 3.16

8 The melting points of the hydrogen compounds of groups 4, 5, 6 and 7 are shown in Figure 3.17.

   a) Explain why H_2O, NH_3 and HF have much higher melting points than expected (when compared to other hydrogen compounds of similar mass).

   b) Describe the **intermolecular bonding** in

     i) PH_3

     ii) H_2S.

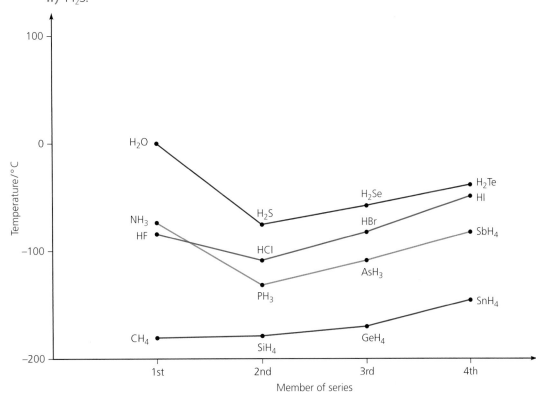

Figure 3.17

9 Water, H_2O, has a boiling point of 100 °C whereas hydrogen sulfide, H_2S, has a boiling point of −60 °C. By referring to the intermolecular forces involved, explain this difference in boiling point.

*10 A student writes the following two statements. Both are *incorrect*. In each case explain the mistake in the student's reasoning.

   a) All ionic compounds are solids at room temperature. Many covalent compounds are gases at room temperature. This proves that ionic bonds are stronger than covalent bonds.

   b) The formula for magnesium chloride is $MgCl_2$ because, in solid magnesium chloride, each magnesium ion is bonded to two chloride ions.

Chapter 4
Oxidising and reducing agents

What you should know

★ Reduction is a gain of electrons by a reactant. Oxidation is a loss of electrons by a reactant.
★ In a redox reaction, reduction and oxidation take place at the same time.
★ An oxidising agent is a substance that accepts electrons. A reducing agent is a substance that donates electrons.
★ Elements with low electronegativities tend to form ions by losing electrons and so act as reducing agents.
★ Elements with high electronegativities tend to form ions by gaining electrons and so act as oxidising agents.
★ In the Periodic Table, the strongest reducing agents are in group 1, and the strongest oxidising agents are in group 7.
★ Compounds, group ions and molecules can act as oxidising or reducing agents:
    ★ Hydrogen peroxide is a molecule that is an oxidising agent.
    ★ Dichromate and permanganate ions are group ions that are strong oxidising agents in acidic solutions.
    ★ Carbon monoxide is a gas that can be used as a reducing agent.
★ Oxidising agents are widely used because of the effectiveness with which they can kill fungi and bacteria, and can inactivate viruses. The oxidation process is also an effective means of breaking down coloured compounds, making oxidising agents ideal for use as 'bleach' for clothes and hair.
★ The electrochemical series represents a series of reduction reactions.
★ The strongest oxidising agents are at the bottom of the left-hand column of the electrochemical series.
★ The strongest reducing agents are at the top of the right-hand column of the electrochemical series.
★ An ion-electron equation can be balanced by adding appropriate numbers of water molecules, hydrogen ions and electrons.
★ Ion-electron equations can be combined to produce redox equations.

Redox reactions

When zinc metal is added to a solution of copper (II) sulfate, a **redox reaction** takes place.

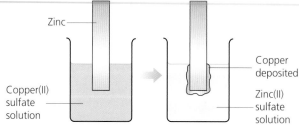

Figure 4.1 An example of a redox reaction

The **electrochemical series** can be used to write ion–electron equations for oxidation and reduction reactions. In the case of the reaction illustrated in Figure 4.1, the zinc metal is higher in the electrochemical series than the copper ions. As a result, the zinc atoms lose electrons and the copper ions gain electrons. Loss of electrons is known as oxidation so we say that the zinc atoms have been oxidised. Gain of electrons is known as reduction so we can say that the copper ions have been reduced. An overall redox reaction can be written by cancelling the electrons and combining the equations:

Oxidation: $Zn(s) \rightarrow Zn^{2+}(aq) + 2e^-$

Reduction: $Cu^{2+}(aq) + 2e^- \rightarrow Cu(s)$

Redox: $Zn(s) + Cu^{2+}(aq) \rightarrow Zn^{2+}(aq) + Cu(s)$

Zinc is acting as a **reducing agent** since it supplies the electrons which cause the copper ions to be reduced.

Cu^{2+} is acting as an **oxidising agent** since it accepts the electrons from the zinc.

> ### Hints & tips
>
> The ion electron equations for writing redox equations are found on page 12 of the data booklet. All the equations are written as reductions, i.e. the ions gain electrons. To write an oxidation equation, find the relevant equation from the data booklet and write it in reverse.

Remember

An oxidising agent is a substance that accepts electrons.

A reducing agent is a substance that donates electrons.

Using electronegativity

Electronegativity values can be used to predict whether a substance will act as an oxidising or reducing agent.

Most metals have low electronegativity values. They act as reducing agents.

Non-metals have higher electronegativity values. They act as oxidising agents.

In the Periodic Table, the strongest reducing agents are in group 1 (the alkali metals). They readily form ions by losing their loosely held outer electron. This electron is transferred to the substance reacting with the alkali metal causing the reacting substance to be reduced. The alkali metals have the lowest electronegativity values of all the elements.

In the Periodic Table, the strongest oxidising agents are in group 7 (the halogens). They readily gain an electron to form a negatively charged ion. They gain this electron from the substance reacting with the halogen, i.e. the halogen causes the substance to lose electrons (it is oxidised). The halogens have very high electronegativity values, e.g. fluorine has the highest value, 4.0, of all elements.

Remember

Reducing agents have low electronegativity values.

Oxidising agents have high electronegativity values.

Examples

The following question appears in various forms in SQA past papers. This version is taken from the 2007 Higher Paper.

★ **Which of the following is a redox reaction?**

A $Mg + 2HCl \rightarrow MgCl_2 + H_2$
B $MgO + 2HCl \rightarrow MgCl_2 + H_2O$
C $MgCO_3 + 2HCl \rightarrow MgCl_2 + H_2O + CO_2$
D $Mg(OH)_2 + 2HCl \rightarrow MgCl_2 + 2H_2O$

Solution

These questions are straightforward to solve. Always look for the equation where the metal is changing from an element to a compound or vice versa. For a metal to form a compound it has to form ions, in other words, it has to oxidise. In this case, the only answer suitable is **A**. The magnesium is oxidised to form Mg^{2+} ions. The reduction equation is the formation of hydrogen gas:

$2H^+ + 2e^- \rightarrow H_2$

In all of the other answers, the magnesium is an ion as a reactant and an ion as a product – it does not change.

A similar question type is shown here. This question came from a Higher Chemistry exam paper in 2008.

★ **In which of the following reactions is the hydrogen ion acting as an oxidising agent?**

A $Mg + 2HCl \rightarrow MgCl_2 + H_2$
B $NaOH + HNO_3 \rightarrow NaNO_3 + H_2O$
C $CuCO_3 + H_2SO_4 \rightarrow CuSO_4 + H_2O + CO_2$
D $CH_3COONa + HCl \rightarrow NaCl + CH_3COOH$

Solution

Again, look for an equation involving a metal changing. The only equation where the metal changes is equation **A** in which the Mg atom turns into Mg ions; it is oxidised. Thus, the hydrogen ions (from the HCl) must be acting as an oxidising agent.

Writing redox equations

Simple redox equations can be written for redox reactions provided it is known which substance is being oxidised and which substance is being reduced. In order to combine two ion–electron equations, the number of electrons must be equal. This is shown in the following examples.

Example

Fluorine gas can be used to oxidise lithium metal. Write the ion electron equations for these reactions and combine them to form an overall redox equation.

Solution

First, identify the relevant equations from the data booklet.

They are listed as:

1 $Li^+ + e^- \rightarrow Li$
2 $F_2 + 2e^- \rightarrow 2F^-$

Second, reverse the lithium equation since you are told that the reaction is with lithium metal (atoms) and that the lithium is being oxidised.

1 $Li \rightarrow Li^+ + e^-$
2 $F_2 + 2e^- \rightarrow 2F^-$

Now, you must ensure that the number of electrons lost by the lithium is the same as that gained by the fluorine. This can be done by multiplying equation 1 by 2.

1 $2Li \rightarrow 2Li^+ + 2e^-$
2 $F_2 + 2e^- \rightarrow 2F^-$

Finally, add both equations together to form an overall redox equation:

$2Li + F_2 \rightarrow 2Li^+ + 2F^-$

Example

Write ion–electron equations and combine them to form a redox equation for the reaction of the permanganate ion with iron (II) ions.

Solution

Since the question states that the reaction is between permanganate and iron (II), the ion–electron equations must start with these ions. Both ion electron equations can be obtained from the ECS:

$MnO_4^- + 8H^+ + 5e^- \rightarrow Mn^{2+} + 4H_2O$

The permanganate equation is, therefore, a reduction which means that we are looking for an ion–electron equation for the *oxidation* of iron (II).

$Fe^{2+} \rightarrow Fe^{3+} + e^-$

The iron (II) equation must be multiplied by 5 so that the number of electrons lost is equal to the number of electrons gained by the permanganate ion.

$MnO_4^- + 8H^+ + 5e^- \rightarrow Mn^{2+} + 4H_2O$

$5Fe^{2+} \rightarrow 5Fe^{3+} + 5e^-$

$MnO_4^- + 8H^+ + 5Fe^{2+} \rightarrow Mn^{2+} + 4H_2O + 5Fe^{3+}$

Writing more complex ion–electron equations

Example

Write an ion–electron equation for the conversion of dichromate into chromium (III) ions.

Solution

Since the ion–electron equation for this reaction is in the data booklet, we will be able to check our answer. The steps for solving these more complex equations are:

1 Check that the main element reacting (not oxygen) is balanced.
2 Add water to balance oxygen atoms.
3 Add H^+ ions to balance the hydrogen atoms.
4 Add electrons to the same side as the H^+ ions so that both sides of the equation have the same charge.

The starting equation is:

$$Cr_2O_7^{2-} \rightarrow Cr^{3+}$$

Step 1: We must double the chromium (III) to balance both sides.

$$Cr_2O_7^{2-} \rightarrow 2Cr^{3+}$$

Step 2: We must add water to the right-hand side to balance the oxygen atoms.

$$Cr_2O_7^{2-} \rightarrow 2Cr^{3+} + 7H_2O$$

Step 3: We must add H^+ ions to the left-hand side to balance the hydrogens.

$$Cr_2O_7^{2-} + 14H^+ \rightarrow 2Cr^{3+} + 7H_2O$$

Step 4: We must add electrons to the left-hand side to ensure the charge is the same on both sides. Since the charge is 12+ on the reactant side and 6+ on the product side, $6e^-$ must be added to the reactant side to give an overall 6+ charge to the reactants. This is the final stage and will give us our solution:

$$Cr_2O_7^{2-} + 14H^+ + 6e^- \rightarrow 2Cr^{3+} + 7H_2O$$

Example

Write an ion-electron equation for the conversion of MnO_4^- into Mn^{2+}.

Solution

Applying the rules shown in the previous example:

1 The Mn is already balanced.
2 Need to add 4 water molecules to the right-hand side to balance the 4 O atoms:
$$MnO_4^- \rightarrow Mn^{2+} + 4H_2O$$
3 Need to add $8H^+$ ions to the left-hand side to balance the 8H atoms on the right:
$$MnO_4^- + 8H^+ \rightarrow Mn^{2+} + 4H_2O$$
4 Need to add $5e^-$ to the left-hand side to give an overall charge of 2+ (as shown on the right):
$$MnO_4^- + 8H^+ + 5e^- \rightarrow Mn^{2+} + 4H_2O$$

Compounds as oxidising and reducing agents

The previous examples illustrate the fact that compounds such as permanganate (MnO_4^-) can be used as oxidising agents. Many compounds can act as either oxidising or reducing agents. For example:

- The dichromate ion ($Cr_2O_7^{2-}$) is a powerful oxidising agent that you will use in practical experiments to oxidise alcohols in Section 2.
- The gas carbon monoxide (CO) is a powerful reducing agent that can be used to extract metals from their compounds, i.e. it can supply the electrons to metal ions to change them into metal atoms.

An examination of the data booklet shows that dichromate and permanganate ions are located at the bottom of the electrochemical series, i.e. strong oxidising agents are found at the bottom of the electrochemical series.

Remember

Elements and compounds at the bottom left of the electrochemical series are strong oxidising agents.

Elements and compounds at the top right of the electrochemical series are strong reducing agents.

Everyday uses for strong oxidising agents

Hydrogen peroxide is a highly effective bleach as it is able to break down coloured compounds. It is found in teeth-whitening products and hair products. Like many oxidising agents, hydrogen peroxide is also an effective antiseptic as it can kill bacteria and fungi and destroy viruses. Potassium permanganate is also used for its antiseptic properties and is commonly used in aquaria to destroy the bacteria and fungi that can infect fish.

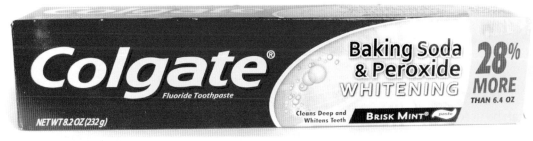

Figure 4.2 Hydrogen peroxide is a powerful oxidising agent used in teeth whitening.

33

Key points

* Ion–electron equations can be written for reactions and such equations can be combined to form redox equations.
* An oxidising agent is a substance that accepts electrons. A reducing agent is a substance that donates electrons.
* Metals tend to act as reducing agents as they readily lose electrons.
* Non-metals tend to act as oxidising agents as they readily gain electrons.
* The electronegativity scale can be used to assess whether a substance is likely to be an oxidising or reducing agent. Reducing agents have low electronegativity values. Oxidising agents have high electronegativity values.
* In the Periodic Table, the strongest reducing agents are in group 1, and the strongest oxidising agents are in group 7.
* In the electrochemical series, the strongest reducing agents are found on the top right, and the strongest oxidising agents are found on the bottom left.
* Everyday applications of oxidising agents include the use of hydrogen peroxide to bleach clothes, hair and teeth. Strong oxidising agents, such as potassium permanganate and hydrogen peroxide are also powerful antiseptics (they destroy bacteria, viruses and fungi).

Study questions

1 Magnesium reacts with silver (I) ions to form silver and a solution of magnesium ions.
   a) Write ion–electron equations for the oxidation and reduction reactions taking place.
   b) Combine the equations to form a redox equation.
   c) Identify the oxidising agent and the reducing agent.

2 Place the following substances in order of their ability to act as an oxidising agent (strongest to weakest): zinc, chlorine, fluorine and sodium.

3 Complete the ion–electron equations for the following:
   a) $SO_3^{2-} \rightarrow SO_4^{2-}$ c) $H_2O_2 \rightarrow O_2$
   b) $NO_3^- \rightarrow NO$ d) $VO_3^- \rightarrow V^{2+}$

4 In which reaction is hydrogen gas acting as a reducing agent?

   A $H_2 + \frac{1}{2} O_2 \rightarrow H_2O$ C $H_2 + 2Li \rightarrow 2LiH$

   B $H_2 + PbO \rightarrow Pb + H_2O$ D $H_2 + C_4H_8 \rightarrow C_4H_{10}$

5 Which of the following is the strongest reducing agent?
   A Chlorine   C Magnesium
   B Sodium   D Iodine

6 Ethanol can be oxidised to form ethanoic acid according to the equation below.
   Complete the ion electron equation for this reaction.
   $CH_3CH_2OH \rightarrow CH_3COOH$

7 Which of the following ions could be used to oxidise silver atoms to form silver ions $Ag \rightarrow Ag^+ + e^-$?
   A MnO_4^- C I_2
   B Mn^{2+} D $2I^-$

8 Carbon monoxide can be used to extract metals from their ores. For example, when iron (III) oxide is reacted with carbon monoxide, iron metal and carbon dioxide are formed.
   a) Write a balanced chemical equation for this reaction.
   b) Write an ion–electron equation for iron (III) forming iron metal.
   c) Hence, state whether carbon monoxide is acting as an oxidising or reducing agent.

Section 2 Nature's Chemistry

Chapter 5
Systematic carbon chemistry

What you should know

★ Compounds containing only single carbon–carbon bonds are described as saturated.
★ Compounds containing at least one carbon–carbon double bond are described as unsaturated.
★ Compounds containing carbon–carbon double bonds can take part in addition reactions. In an addition reaction, two molecules combine to form a single molecule.
★ It is possible to distinguish an unsaturated compound from a saturated compound using bromine solution. Unsaturated compounds quickly decolourise bromine solution.
★ The structure of any molecule can be drawn as a full or a shortened structural formula.
★ Isomers:
    ★ are compounds with the same molecular formula but different structural formulae
    ★ may belong to different homologous series
    ★ usually have different physical properties.
★ Given the name or a structural formula for a compound, an isomer can be drawn. Isomers can be drawn for a given molecular formula.
★ The solubility, boiling point and volatility (ease of evaporation) of a compound can be predicted by considering
    ★ the presence of O–H or N–H bonds, which implies hydrogen bonding
    ★ the spatial arrangement of polar covalent bonds which could result in a molecule possessing a permanent dipole
    ★ molecular size which would affect London dispersion forces
    ★ the polarities of solute and solvent. Polar or ionic compounds tend to be soluble in polar solvents, non-polar compounds tend to be soluble in non-polar solvents.
★ Solubility, boiling point and volatility can be explained in terms of the type and strength of intermolecular forces present.

Carbon compounds

In Section 2, you will encounter many carbon compounds that are used in medicines, cosmetics and the food industry. This chapter is designed to help you revise previous chemical knowledge of carbon compounds that will be required to help you understand the chemistry of the new compounds you will learn about in this section.

Types of formula

Representation of hydrocarbon compounds can be shown by a **molecular formula**, full **structural formula** and shortened structural formula. This is shown in Table 5.1 which lists the first eight members of the alkane **homologous series**.

Table 5.1 The alkanes

Number of C atoms	Name	Molecular formula	Full structural formula	Shortened structural formula	Boiling point (°C)
1	methane	CH_4	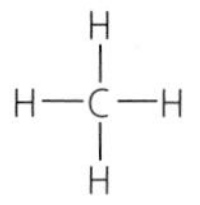	CH_4	−162
2	ethane	C_2H_6	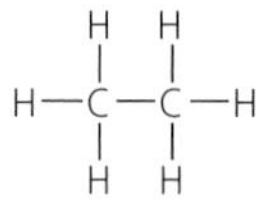	CH_3CH_3	−89
3	propane	C_3H_8	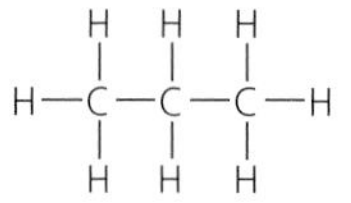	$CH_3CH_2CH_3$	−42
4	butane	C_4H_{10}	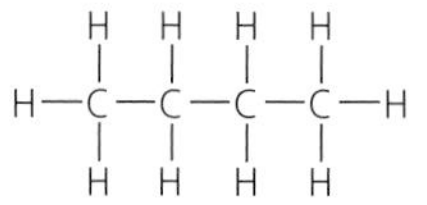	$CH_3CH_2CH_2CH_3$	−1
5	pentane	C_5H_{12}	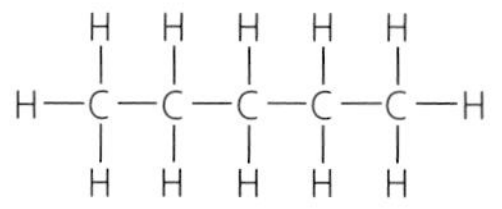	$CH_3CH_2CH_2CH_2CH_3$	36
6	hexane	C_6H_{14}	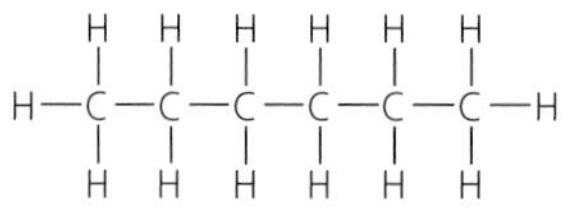	$CH_3CH_2CH_2CH_2CH_2CH_3$	69
7	heptane	C_7H_{16}	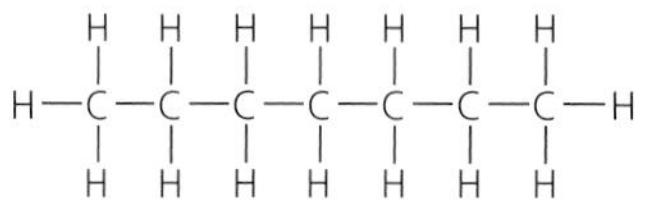	$CH_3CH_2CH_2CH_2CH_2CH_2CH_3$	98
8	octane	C_8H_{18}	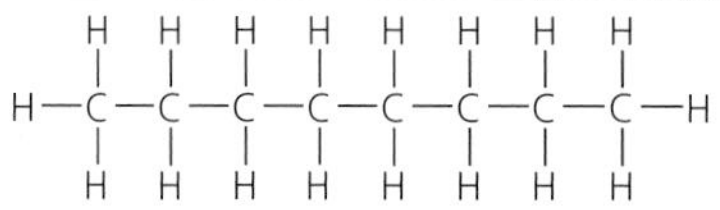	$CH_3CH_2CH_2CH_2CH_2CH_2CH_2CH_3$	126

Similar representations can be shown for the **cycloalkanes** (Table 5.2) and the **alkenes** (Table 5.3).

Table 5.2 The cycloalkanes

Number of C atoms	Name	Molecular formula	Full structural formula	Shortened structural formula	Boiling point (°C)
3	cyclopropane	C_3H_6	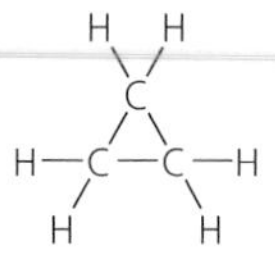	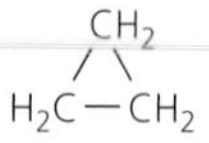	−33
4	cyclobutane	C_4H_8	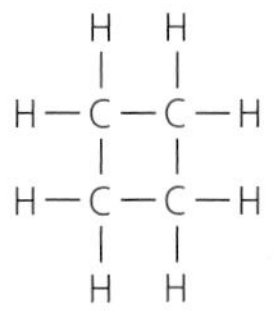	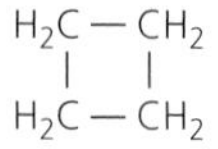	13

Table 5.2 (continued)

Number of C atoms	Name	Molecular formula	Full structural formula	Shortened structural formula	Boiling point (°C)
5	cyclopentane	C_5H_{10}	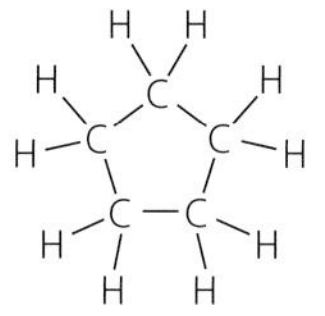	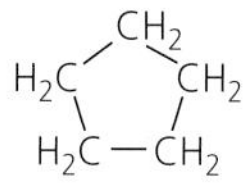	49
6	cyclohexane	C_6H_{12}	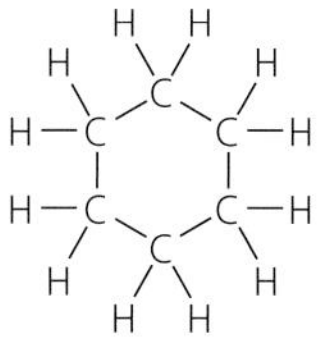	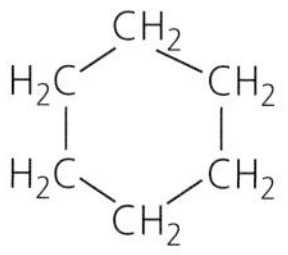	81
7	cycloheptane	C_7H_{14}	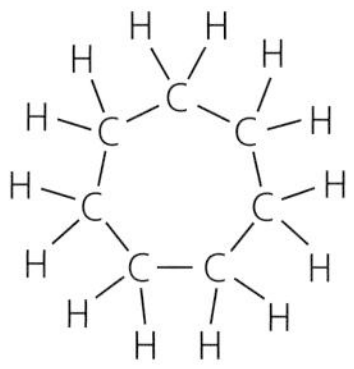	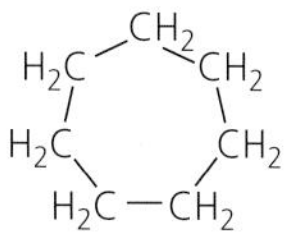	118
8	cyclooctane	C_8H_{16}	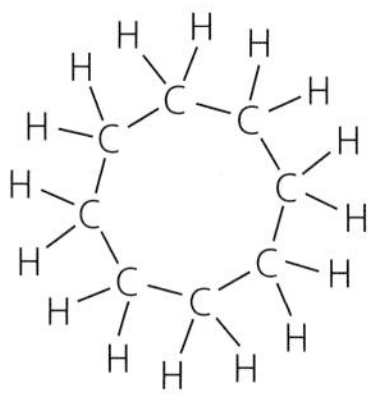	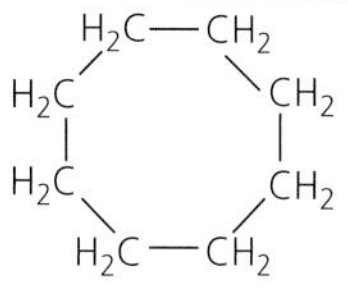	149

Table 5.3 The alkenes

Number of C atoms	Name	Molecular formula	Full structural formula	Shortened structural formula	Boiling point (°C)
2	ethene	C_2H_4	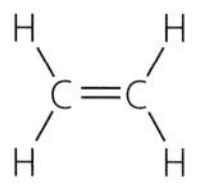	$CH_2{=}CH_2$	−104
3	propene	C_3H_6	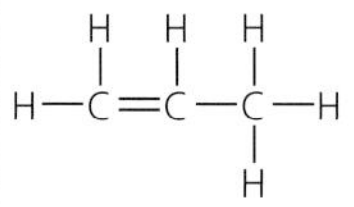	$CH_2{=}CHCH_3$	−48
4	butene	C_4H_8	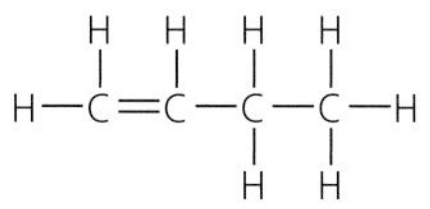	$CH_2{=}CHCH_2CH_3$	−6
5	pentene	C_5H_{10}	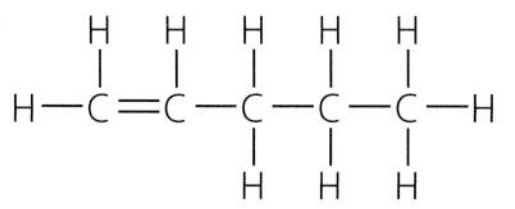	$CH_2{=}CHCH_2CH_2CH_3$	30
6	hexene	C_6H_{12}	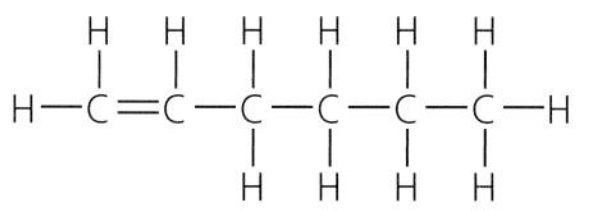	$CH_2{=}CHCH_2CH_2CH_2CH_3$	63
7	heptene	C_7H_{14}	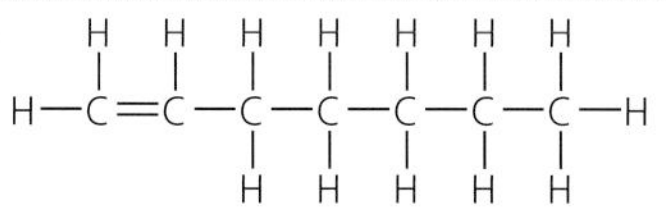	$CH_2{=}CHCH_2CH_2CH_2CH_2CH_3$	115
8	octene	C_8H_{16}	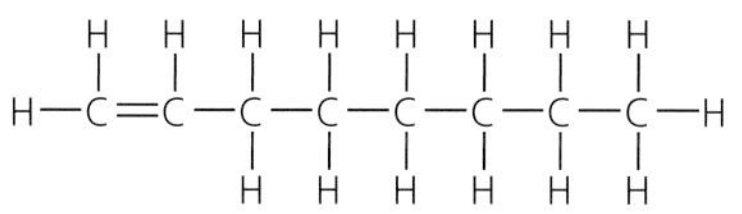	$CH_2{=}CHCH_2CH_2CH_2CH_2CH_2CH_3$	122

Compounds containing only single carbon–carbon bonds, like the alkanes and cycloalkanes, are described as **saturated**.

Compounds containing at least one carbon–carbon double bond, like the alkenes, are described as **unsaturated**.

Boiling points, volatility and solubility

An examination of the boiling points for all three families of hydrocarbons shows us the same trend, i.e. the boiling points increase as the molecules get bigger in size. This can be explained by the fact that the main force of attraction between hydrocarbon molecules is London dispersion forces. The bigger molecules have more electrons than the smaller molecules, so the LDF between bigger molecules is stronger. Hence, it takes more energy to break apart the intermolecular forces (LDF) between bigger molecules than smaller molecules.

This also explains why the smaller molecules are more volatile, i.e. they evaporate more easily. Evaporation involves molecules escaping from the liquid phase and becoming free molecules in the gas phase. If you take liquid cyclopentane, for example, you can smell the cyclopentane vapour at room temperature as the temperature of the room is high enough to cause a significant number of the cyclopentane molecules to break free from the liquid phase. The temperature of the room is enough to overcome the weak forces (LDF) holding some of the molecules together in the liquid phase.

As the hydrocarbons are non-polar molecules, they do not dissolve in water as they cannot form hydrogen bonds to water molecules. Likewise, they cannot dissolve in other polar compounds such as ethanol.

As they are non-polar, they are useful solvents for dissolving other non-polar substances. Cyclohexane and hexane are routinely used as non-polar solvents. Another advantage of using these hydrocarbons as solvents is that they evaporate easily. This means that they can be used to extract compounds (e.g. an oil from a food) and can then be evaporated to leave behind the desired compound.

Isomers

Compounds with the same molecular formula but different structural formula are known as **isomers**. For example, propene and cyclopropane both have the molecular formula C_3H_6 but have different structures, as can be seen from Tables 5.2 and 5.3. Isomers also have different physical and chemical properties. You will note that the boiling points are different and will recall that, as propene is unsaturated, propene can undergo addition reactions but cyclopropane cannot.

Remember

You should be able to use the molecular, full structural and shortened structural formula for drawing carbon compounds.

Remember

The intermolecular force between hydrocarbon molecules is LDF.

Isomers of saturated compounds are usually formed by forming branches, as shown by the representation of two isomers of butane in Figure 5.1.

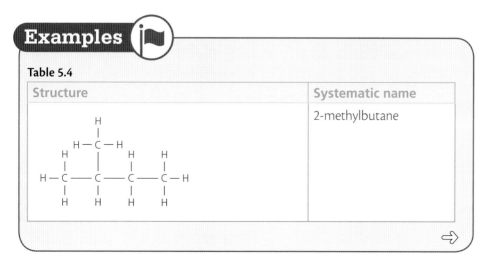

Figure 5.1 Structures with the formula C_4H_{10}

A useful way of identifying isomers is to use systematic naming since compounds with the same formula but a different name will certainly be isomers. For example, the two compounds in Figure 5.1 would be named butane and 2-methylpropane.

A reminder of the rules for systematically naming compounds is shown below.

Systematic naming

Rule 1 Identify the parent compound with the longest chain of carbon atoms and name the alkane, cycloalkane or alkene with this number of carbon atoms.
Rule 2 For alkenes, number from the side of the molecule that gives the double bond the lowest number.
Rule 3 For branches, name them according to the number of carbon atoms (methyl, ethyl, etc.) ensuring that you number from the side of the molecule that gives the branch the lowest number.
Rule 4 Where there are branches and double bonds, number from the side that gives the double bond the lowest number.
Rule 5 Use the terms di-, tri- , etc., when describing two or three double bonds or branches of the same type.

Examples

Table 5.4

Structure	Systematic name
	2-methylbutane

⇒ **Table 5.4** (continued)

Structure	Systematic name
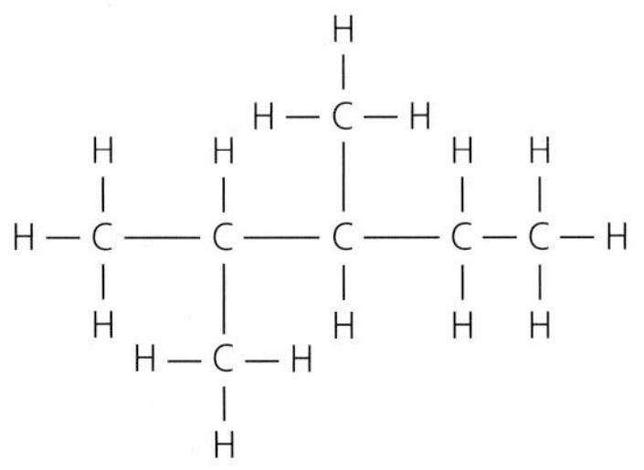 (structure of 2,3-dimethylpentane)	2,3-dimethylpentane
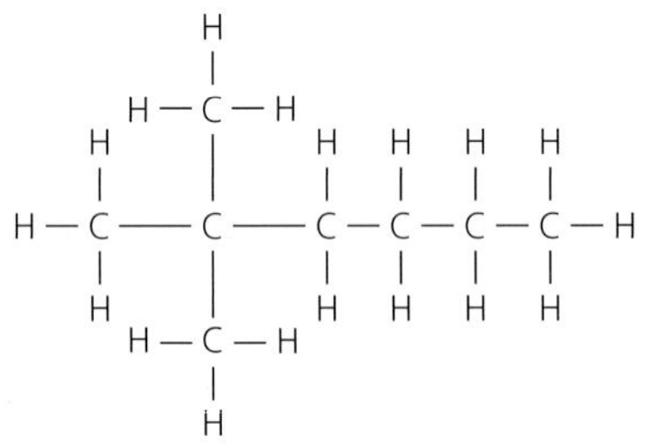 (structure of 2,2-dimethylhexane)	2,2-dimethylhexane
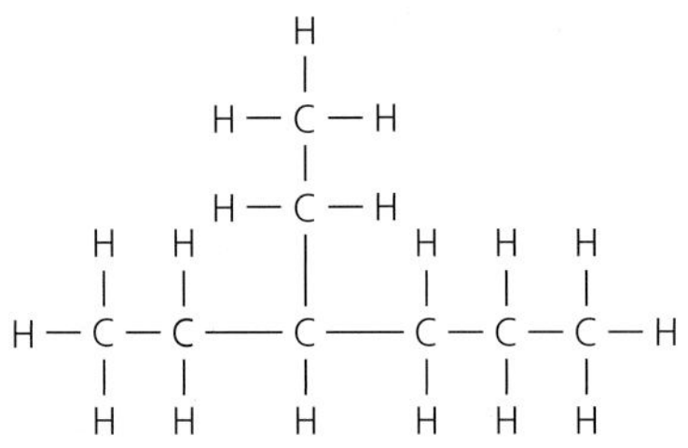 (structure of 3-ethylhexane)	3-ethylhexane
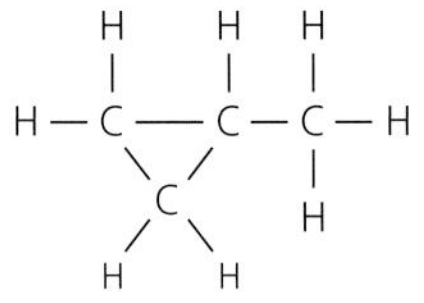 (structure of methylcyclopropane)	methylcyclopropane
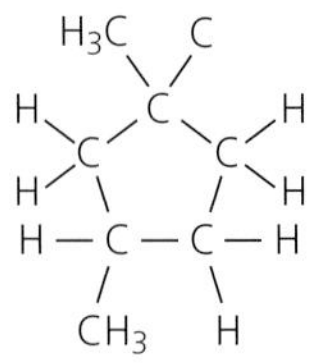 (structure of 1,3-dimethylcyclopentane)	1,3-dimethylcyclopentane
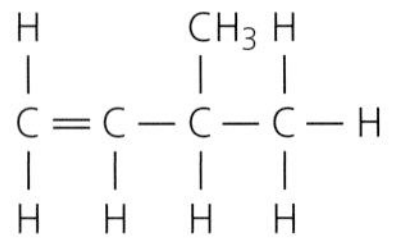 (structure of 3-methylbut-1-ene)	3-methylbut-1-ene
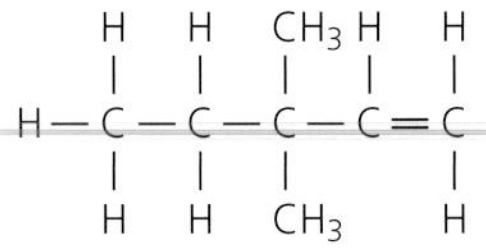 (structure of 3,3-dimethylpent-1-ene)	3,3-dimethylpent-1-ene
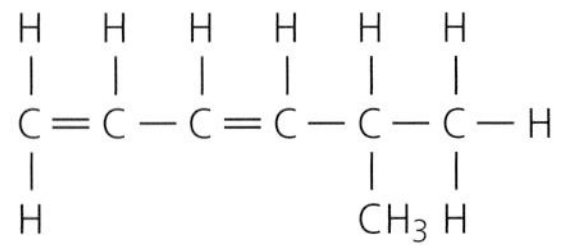 (structure of 5-methylhex-1,3-diene)	5-methylhex-1,3-diene

Addition reactions

Compounds containing double bonds can take part in **addition** reactions. This fact allows bromine solution to be used to distinguish between saturated and unsaturated compounds since unsaturated compounds will rapidly decolourise bromine water.

Some examples of common addition reactions are shown below.

Hints & tips

When drawing carbon compounds, always re-check your structure to make sure each carbon atom has 4 bonds. If not, you know you have made a mistake!

Examples

Table 5.5

(a) Reaction of propene with bromine to form 1,2-dibromopropane	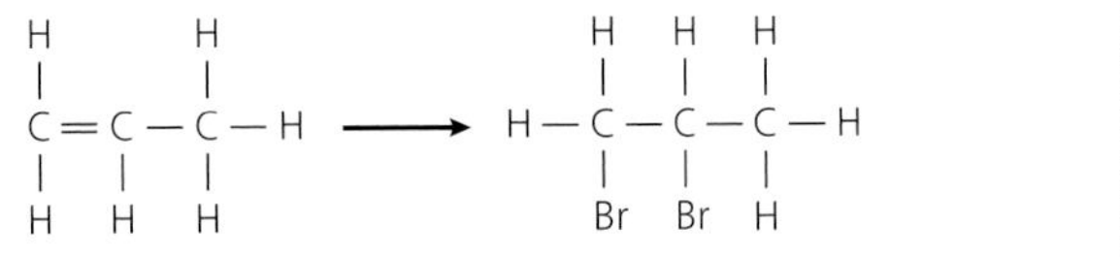 H H / C=C—C—H with H H H → H—C—C—C—H with Br Br H
(b) Reaction of ethene with hydrogen to form ethane	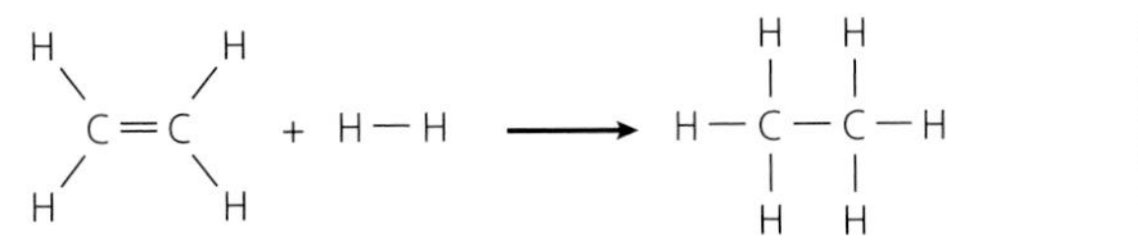 H₂C=CH₂ + H—H → H—C—C—H
(c) Reaction of ethene with chlorine to form 1,2-dichloroethane	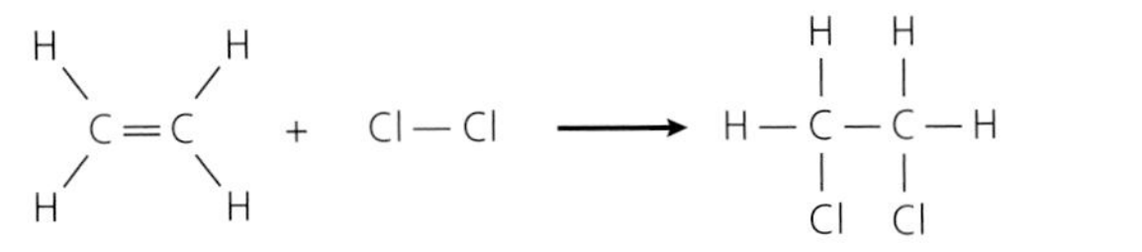 H₂C=CH₂ + Cl—Cl → H—C—C—H with Cl Cl
(d) Reaction of propene with HCl. Note that there are two possible products: 1-chloropropane and 2-chloropropane.	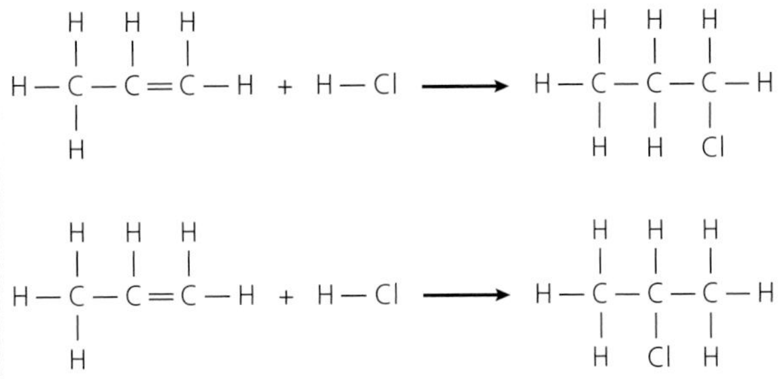 H—C—C=C—H + H—Cl → H—C—C—C—H (H H Cl) ; H—C—C=C—H + H—Cl → H—C—C—C—H (H Cl H)
(e) Reaction of propene with water. Again, note that there are two possible products: propan-1-ol and propan-2-ol.	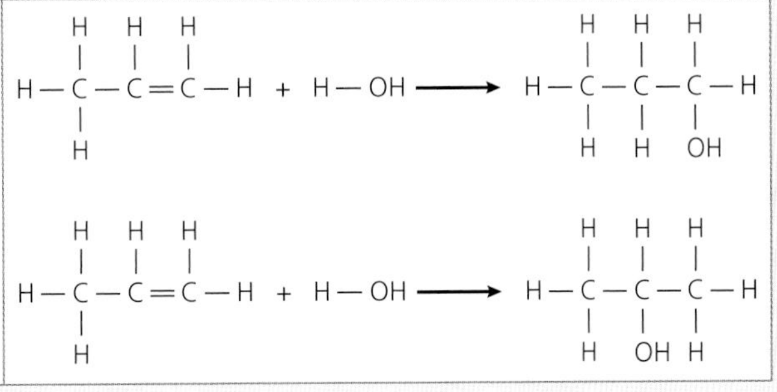 H—C—C=C—H + H—OH → H—C—C—C—H (H H OH) ; H—C—C=C—H + H—OH → H—C—C—C—H (H OH H)

Hints & tips

The reaction between an unsaturated compound and Br_2 is the same for all halogens (F_2, Cl_2, I_2). Just replace Br with the halogen being added. Likewise for adding HCl, the reaction is the same for other hydrogen halides (HF, HBr and HI): replace the Cl with the halogen atom being added.

Key points

* Hydrocarbons can be named using systematic naming rules.
* Isomers are compounds with the same molecular formula but different structural formula. They usually have different chemical and physical properties.
* Structures with only single carbon–carbon bonds are described as saturated. Structures with at least one carbon–carbon double bond are described as unsaturated.
* Unsaturated compounds take part in addition reactions with halogens, hydrogen halides (e.g. HCl and HBr), hydrogen and water.
* Compounds that rapidly decolourise bromine solution are unsaturated.
* Hydrocarbons have low boiling points, evaporate easily, and do not dissolve in water because they are non-polar compounds held together by weak LDF. As the molecules increase in size, the LDF between molecules increases leading to higher boiling points.

Study questions

1 Use the grid below to answer the questions that follow.

Table 5.6

A	B
H H | | H — C — C — H | | H — C — C — H | | H H	H | H — C — H H | H | | | H — C —— C —— C — H | | | H H H

C	D
H H H H | | | | H — C — C — C — C — Cl | | | | H H H H	H H H H | | | | H — C — C — C — C — H | | | | H Cl H H

a) Which compound is an isomer of butene?
b) Which compound could be formed by the reaction of but-2-ene and HCl?
c) State the systematic name of the compound shown in B.
d) Name two reactants that could be used to form the compound shown in C.

2 Draw full structural formula for the following compounds:
a) 2,4-dimethylhexane
b) 2,2-dimethylpentane
c) 1,2-dimethylcyclopentane
d) 1,2-dimethylcyclobutane
e) 2-methylbut-1-ene
f) 2,3-dimethylbut-1-ene
g) 4,5-dimethylhex-2-ene
h) 4-methylpent-1,3-diene

3 Draw full structural formula for compounds A–E in the following reactions:

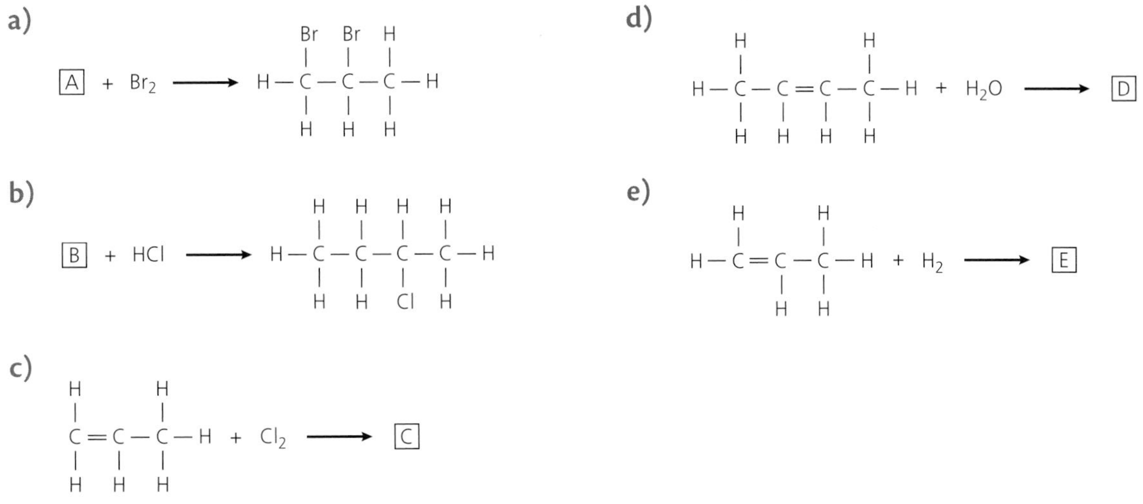

a)

$$A + Br_2 \longrightarrow$$

```
      Br  Br  H
      |   |   |
  H—C — C — C—H
      |   |   |
      H   H   H
```

d)

```
      H           H
      |           |
  H—C — C = C — C—H  + H₂O  ⟶  D
      |   |   |   |
      H   H   H   H
```

b)

$$B + HCl \longrightarrow$$

```
      H   H   H   H
      |   |   |   |
  H—C — C — C — C—H
      |   |   |   |
      H   H   Cl  H
```

e)

```
      H       H
      |       |
  H—C = C — C—H  + H₂  ⟶  E
          |   |
          H   H
```

c)

```
  H       H
  |       |
  C = C — C—H  +  Cl₂  ⟶  C
  |   |   |
  H   H   H
```

Figure 5.2

4 Compound X reacted with HF to form 2-fluorobutane. Name compound X.

5 Compound Y was reacted with water and formed ethanol. Name compound Y.

6 Explain why hexane is insoluble in water but dissolves in cyclohexane.

7 Limonene is a compound found in food that has the following structure:

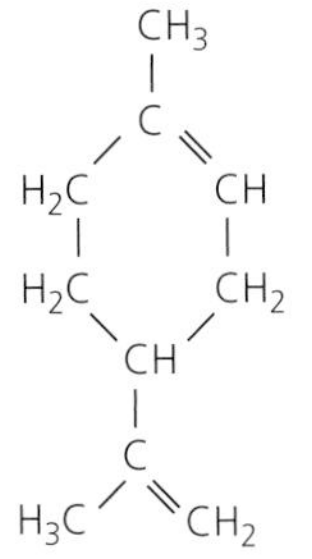

```
              CH₃
              |
           /  C  ⫶
      H₂C       CH
       |        |
      H₂C       CH₂
         \     /
          CH
          |
          C
     H₃C /  ⫶ CH₂
```

Figure 5.3

a) Give an example of a chemical test that could be used to show that a food contains limonene.
b) Suggest a suitable solvent for extracting limonene from food. Explain your answer.
c) Explain why limonene has a higher boiling point than butane.

Chapter 6
Alcohols

What you should know

★ An alcohol is a molecule containing a hydroxyl functional group, −OH group.
★ Straight-chain and branched alcohols can be systematically named, indicating the position of the hydroxyl group from structural formulae containing no more than eight carbon atoms in their longest chain.
★ A molecular formula can be written or a structural formula drawn from the systematic name of a straight-chain or branched alcohol that contains no more than eight carbon atoms in its longest chain.
★ Alcohols can be classified as primary, secondary or tertiary.
★ Alcohols containing two hydroxyl groups are called 'diols' and those containing three hydroxyl groups are called 'triols'.
★ Hydroxyl groups make alcohols polar and this gives rise to hydrogen bonding. Hydrogen bonding can be used to explain the properties of alcohols, including boiling points, melting points, viscosity and solubility/**miscibility** in water.

Naming and classifying alcohols

Alcohols are carbon compounds that contain the hydroxyl **functional group**, −OH. The names of some straight-chain alcohols are shown in Table 6.1.

Table 6.1 Naming straight-chain alcohols

Alkanes	Alcohols
Methane CH_4	Methanol CH_3OH
Ethane C_2H_6	Ethanol C_2H_5OH
Propane C_3H_8	Propanol C_3H_7OH
Butane C_4H_{10}	Butanol C_4H_9OH
Pentane C_5H_{12}	Pentanol $C_5H_{11}OH$
Hexane C_6H_{14}	Hexanol $C_6H_{13}OH$
Heptane C_7H_{16}	Heptanol $C_7H_{15}OH$
Octane C_8H_{18}	Octanol $C_8H_{17}OH$
General formula: C_nH_{2n+2}	General formula: $C_nH_{2n+1}OH$ or $C_nH_{2n+2}O$

Alcohols are named by:

● numbering from the side which gives the −OH the lowest number
● ensuring the −OH takes priority over any branches.
● using the term diol to refer to an alcohol with two hydroxyl groups and triol for an alcohol with three hydroxyl groups.

The alcohol shown in Figure 6.1 would be called 3-methylbutan-1-ol.

In the case of the alcohol in Figure 6.1, numbering of the carbon atoms starts from the right-hand side of the molecule to give the lowest number to the −OH. This places the methyl branch on position 3.

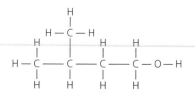

Figure 6.1 Systematic naming of alcohols

Alcohols can be subdivided into three different types depending on the position of the hydroxyl group. These are summarised in Table 6.2.

Table 6.2 Classification of alcohols

Type	Primary	Secondary	Tertiary
	The –OH is attached to a C atom that has at least two H atoms	The –OH is attached to a C atom that has 1 H atom	The –OH is attached to a C atom that is not attached to any H atoms.
Position of –OH group	Joined to the *end* of the carbon chain	Joined to an *intermediate* carbon atom	Joined to an *intermediate* carbon atom which also has a branch attached
Characteristic group of atoms	–CH$_2$OH	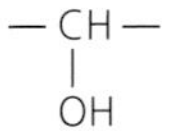 $-CH-$ $\quad\vert$ $\quad$OH	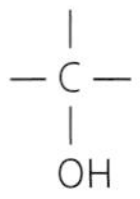 $\quad\vert$ $-C-$ $\quad\vert$ $\quad$OH

Table 6.3 shows the systematic names and classification of some alcohols.

Table 6.3 Naming alcohols

Structure	Name	Classification
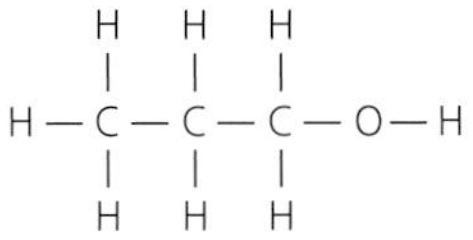	Propan-1-ol	Primary
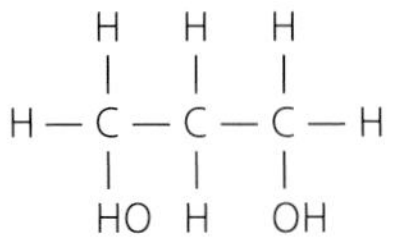	Propan-1,3-diol	Primary
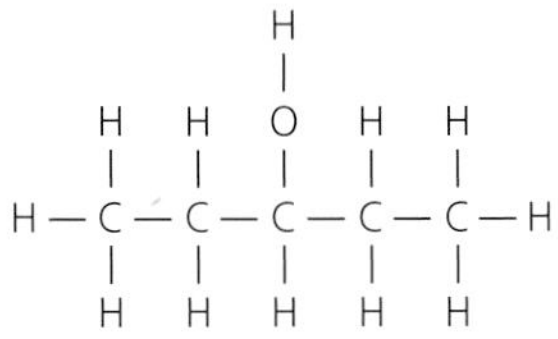	Pentan-3-ol	Secondary
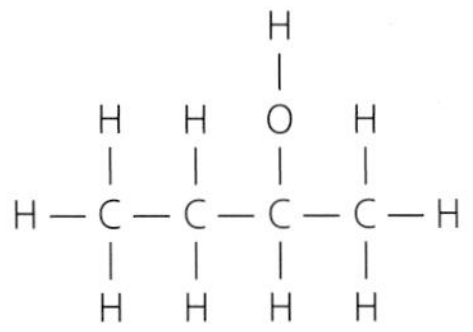	Butan-2-ol	Secondary
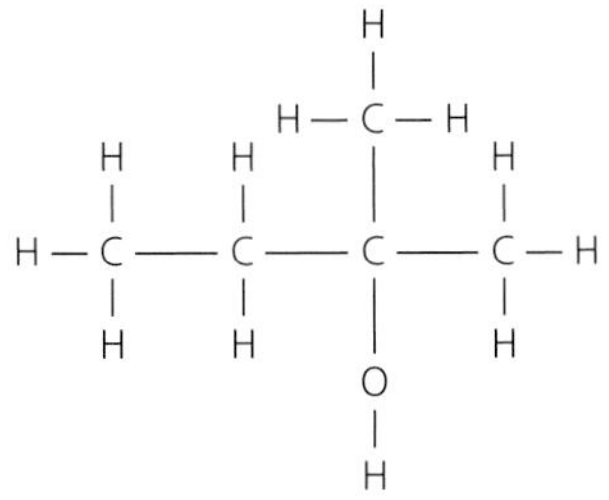	2-Methylbutan-2-ol	Tertiary
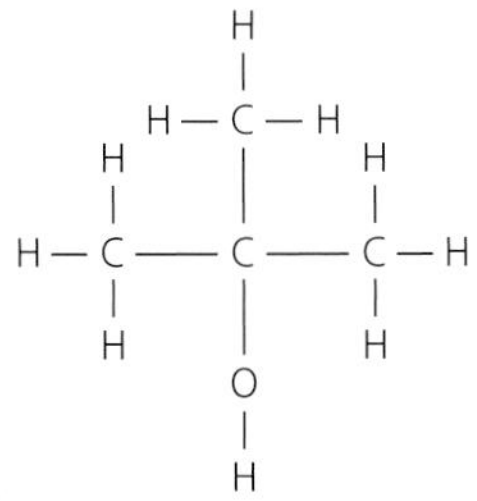	2-Methylpropan-2-ol	Tertiary

Hints & tips

When drawing the structure of an alcohol, remember to draw a bond going from the C atom directly to the O atom and not the H atom since it is the O that bonds to the C and not the H.

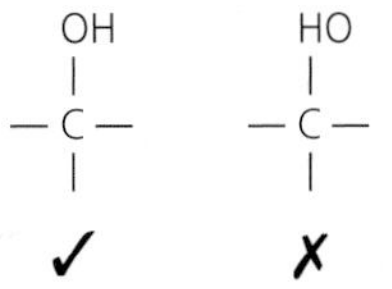

Figure 6.2

Alcohols and hydrogen bonding

Alcohols can form hydrogen bonds because of the polar −OH group. Hydrogen bonding between two ethanol molecules is shown in the diagram below.

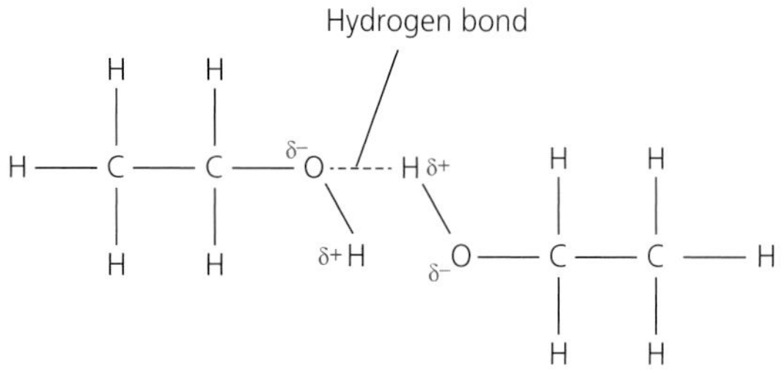

Figure 6.3

An examination of the structures shown in Table 6.4 illustrates the effect of hydrogen bonding. Propane, which does not have a hydroxyl group, cannot form a hydrogen bond with other propane molecules. Consequently, propane has a much lower boiling point than the other molecules as only weak London dispersion forces have to be broken to change propane from a liquid to a gas. Adding hydroxyl groups has a significant effect on the boiling point as the molecules are held together by hydrogen bonding in the solid and liquid state. Much more energy must be supplied to overcome the hydrogen bonds which attract the molecules. Now compare the alcohols: as the number of −OH groups increases, more hydrogen bonds can be formed between molecules. Therefore, the boiling point increases as more energy is required to break the hydrogen bonds.

Remember

Alcohols can form hydrogen bonds due to the presence of the −OH group.

Table 6.4 Alcohols and hydrogen bonding

Name of compound	Structure	Boiling point/°C
Propane	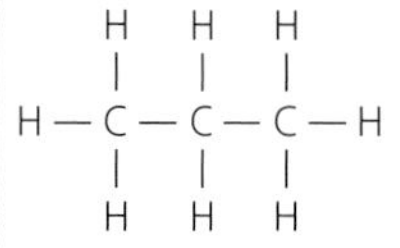	−42
Propan-1-ol	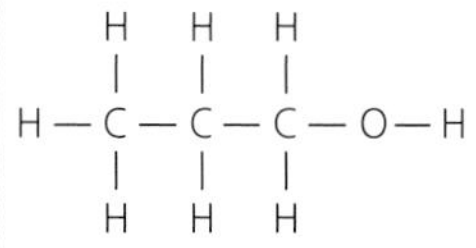	97
Propan-1,2-diol	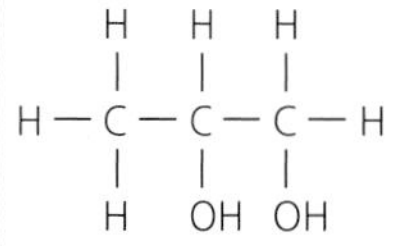	188
Propane-1,2,3-triol (Glycerol)	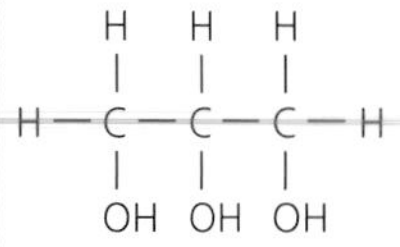	290

Alcohols and viscosity

The experiment shown in Figure 6.4 illustrates the effect of hydrogen bonding on compounds, like alcohols, which contain an −OH group.

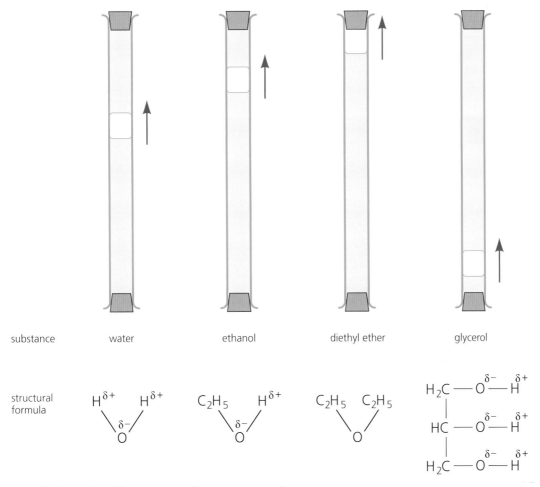

Figure 6.4 The effect of hydrogen bonding on compounds

The experiment involves moving an air bubble through each liquid. This can be used to compare the viscosity of each liquid (the 'thickness' of the liquid), i.e. the longer it takes the air bubble to move from one end to the other, the more viscous the liquid. The diagrams show that glycerol (with three −OH groups) is the most viscous followed by water (two −OH groups), ethanol (one −OH group) and then diethyl ether (no −OH groups; cannot H bond).

The conclusion is: the more hydrogen bonds that can form between molecules, the higher the viscosity.

Alcohols and solubility

As alcohols are polar, they dissolve in other polar liquids such as water. They will not, however, dissolve in non-polar liquids like hexane.

Key points

* Alcohols (compounds containing the hydroxyl group) can be named using rules for systematic naming and can be classified as primary, secondary or tertiary by noting the position of the hydroxyl group.

* Alcohols are polar molecules and can form hydrogen bonds. This explains why they have higher boiling points, higher viscosities and are soluble in polar compounds such as water.

Study questions

1 Name the alcohols listed in the table below and classify them as primary, secondary or tertiary.

Table 6.5

Alcohol	Name	Classification
H—C—C—C—OH (propan-1-ol skeleton with H's)	a)	b)
H—C—C—C—C—C—H with CH₃ and OH substituents	c)	d)
H—C—C—C—C—C—C—H with CH₃, OH, OH	e)	f)
H—C—C—C—C—H with CH₃ and OH	g)	h)
H—C—C—C—C—C—OH with CH₃, CH₃	i)	j)
H—C—C—C—C—C—C—H with OH, CH₃, OH	k)	l)

2 Draw full structural formula for the following alcohols:
   a) 2-methylpentan-1-ol
   b) 2,2-dimethylhexan-1-ol
   c) butan-1,2,3-triol
   d) hexan-2,2-diol
   e) 2-methylbutan-1,2-diol

*3 The following question is taken from SQA Higher Chemistry, 2015, page 14, Q.4d (ii).

Geraniol is a compound found in perfume. It has the following structural formula and systematic name.

Linalool can also be present. Its structural formula is shown.

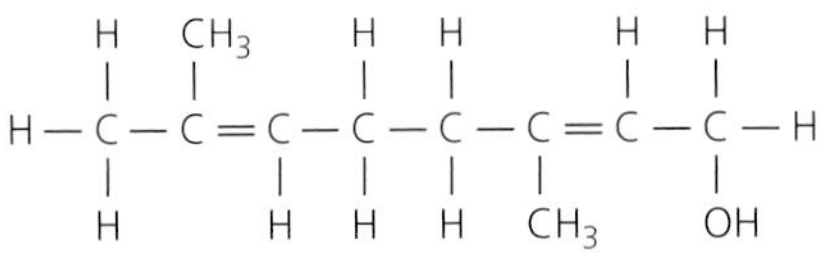

```
      H    CH₃      H   H         H    H
      |    |        |   |         |    |
  H — C — C = C — C — C — C = C — C — H
      |        |   |   |   |         |
      H        H   H   H   CH₃      OH
        3,7-dimethylocta-2,6-dien-1-ol
```

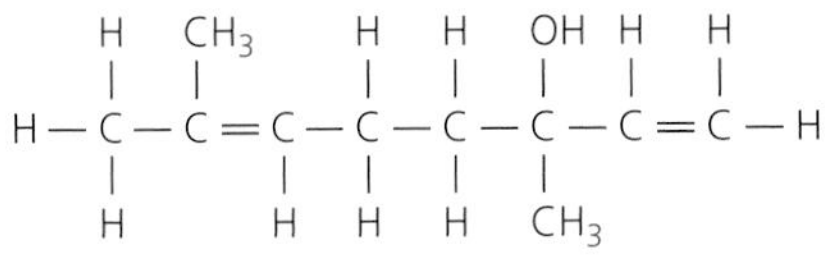

```
      H    CH₃      H   H   OH  H    H
      |    |        |   |   |   |    |
  H — C — C = C — C — C — C — C = C — H
      |        |   |   |   |
      H        H   H   H   CH₃
```

Figure 6.5

a) State the systematic name for linalool.
b) Explain why linalool can be classified as a tertiary alcohol.

4 Ethanol can dissolve in water as it can hydrogen bond to water molecules. Draw full structural formula for water and ethanol and use a dotted line to show how the two molecules can join by hydrogen bonding. Label the atoms involved with the correct charge symbols.

5 Explain why methanol has a higher boiling point than ethane.

6 A student compared the viscosities of propane-1,2,3-triol with propan-1-ol. Explain what you would expect the student to discover from their experiment.

Chapter 7
Carboxylic acids

What you should know

- ★ A carboxylic acid is a molecule containing the carboxyl functional group, –COOH.
- ★ Straight-chain and branched carboxylic acids can be systematically named from structural formulae containing no more than eight carbons in the longest chain.
- ★ A molecular formula can be written or a structural formula drawn from the name of a carboxylic acid that contains no more than eight carbon atoms in its longest chain.
- ★ Carboxylic acids can react with bases:

   metal oxide + carboxylic acid → salt + water
   metal hydroxide + carboxylic acid → salt + water
   metal carbonate + carboxylic acid → salt + water + carbon dioxide

- ★ The name of the salt produced depends on the acid and base used.
- ★ Solubility and melting/boiling point can be explained in terms of the type and strength of intermolecular forces present between carboxylic acid molecules.

Naming carboxylic acids

Carboxylic acids can be identified by the presence of the **carboxyl group** (–COOH), which is shown in Figure 7.1

The names and structures of the first straight-chain carboxylic acids are shown in Table 7.1.

Table 7.1 Names and structures of the first straight-chain carboxylic acids

Name	Full structural formula
methanoic acid	O ‖ H — C — OH
ethanoic acid	H O | ‖ H — C — C — OH | H
propanoic acid	H H O | | ‖ H — C — C — C — OH | | H H
butanoic acid	H H H O | | | ‖ H — C — C — C — C — OH | | | H H H
pentanoic acid	H H H H O | | | | ‖ H — C — C — C — C — C — OH | | | | H H H H

Figure 7.1 Vinegar is a solution of ethanoic acid.

Table 7.1 (continued)

Name	Full structural formula
hexanoic acid	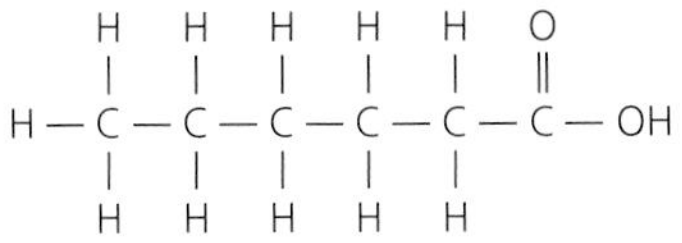 H—C—C—C—C—C—C—OH (with H atoms on carbons and O double bonded to terminal C)
heptanoic acid	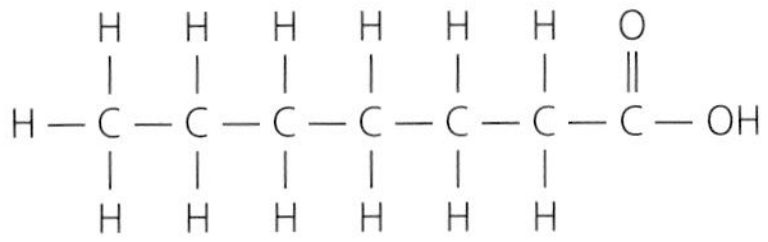 H—C—C—C—C—C—C—C—OH (with H atoms on carbons and O double bonded to terminal C)
octanoic acid	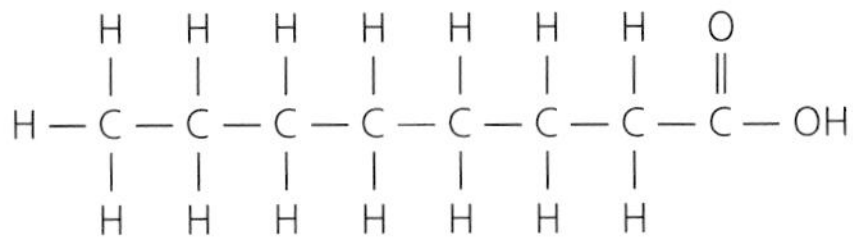 H—C—C—C—C—C—C—C—C—OH (with H atoms on carbons and O double bonded to terminal C)

Branched carboxylic acids are named by giving priority to the carboxyl group. Examples are shown in Table 7.2.

Table 7.2 Branched carboxylic acids

Structure	Name
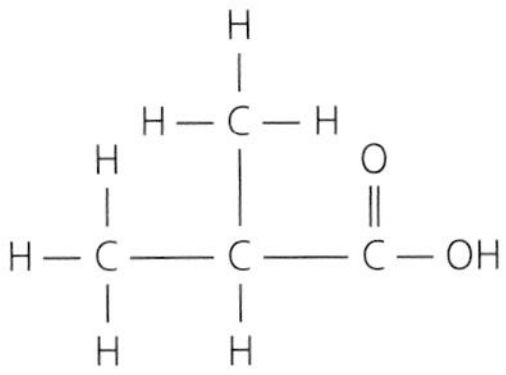 (branched structure)	methylpropanoic acid
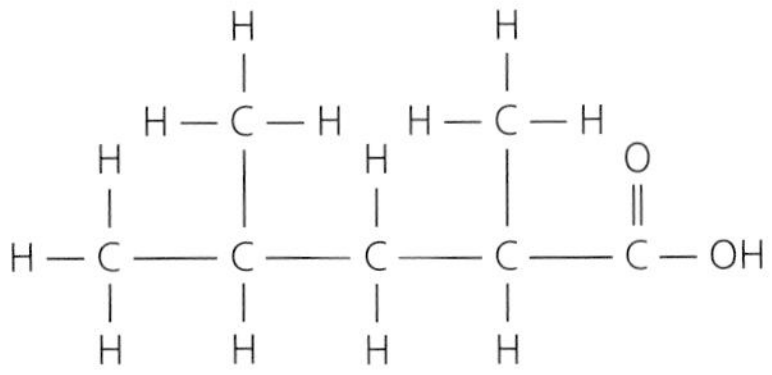 (branched structure)	2,4-dimethylpentanoic acid
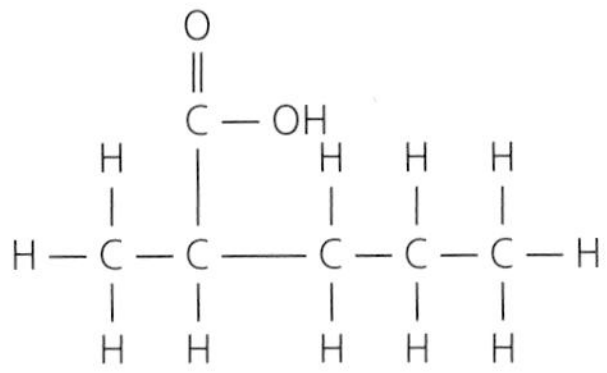 (branched structure)	2-methylpentanoic acid
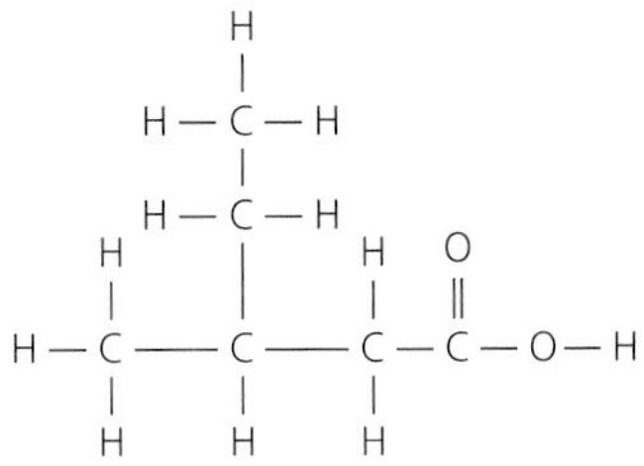 (branched structure)	3-methylpentanoic acid

Reactions of carboxylic acids

Carboxylic acids are acidic because they can dissociate (break up) to form hydrogen ions when added to an aqueous solution (Figure 7.2).

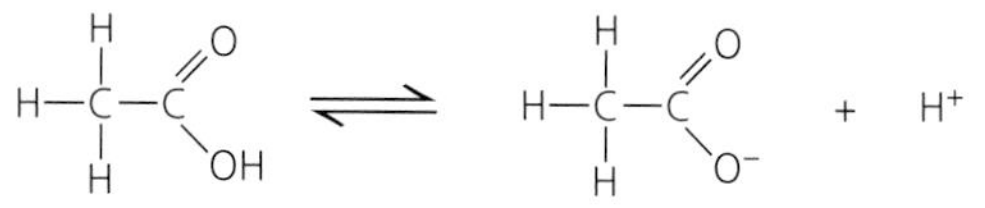

Figure 7.2 Carboxylic acids dissociate to form hydrogen ions in aqueous solution

As they are acids, carboxylic acids can undergo the same types of reaction as the other common laboratory acids such as hydrochloric acid. For example, when sodium hydroxide is added to hydrochloric acid, sodium chloride and water are produced:

$$HCl\ (aq) + NaOH(aq) \rightarrow NaCl(aq) + H_2O(l)$$

The H^+ of the acid is replaced by the Na^+ from the base to form a salt.

Carboxylic acids react to produce salts in the same way. For example:

$$CH_3\ COOH + NaOH \rightarrow CH_3COONa + H_2O$$

ethanoic acid + sodium hydroxide $\rightarrow$ sodium ethanoate + water

> **Hints & tips**
>
> *The salts formed are named by changing the '-oic' ending of the carboxylic acid to '-oate.' For example, the salt formed from methanoic acid would be a methanoate salt. Further examples of salt endings are listed in the following table.*

Table 7.3

Carboxylic acid	Reacting with	Salt
methanoic	sodium hydroxide	sodium methanoate
ethanoic	potassium hydroxide	potassium ethanoate
propanoic	lithium oxide	lithium propanoate
butanoic	magnesium carbonate	magnesium butanoate
pentanoic	ammonium hydroxide	ammonium pentanoate
hexanoic	calcium oxide	calcium hexanoate
heptanoic	sodium carbonate	sodium heptanoate
octanoic	lithium hydroxide	lithium octanoate

The reactions of carboxylic acids studied in Higher Chemistry are the same as those encountered with lab acids in previous chemistry. A list of these reactions, along with an example, is given below. Once you have reviewed this list, try the worked examples which follow.

Reaction with a metal oxide

Table 7.4

General equation	**metal oxide + carboxylic acid → salt + water**
Word equation	lithium oxide + ethanoic acid → lithium ethanoate + water
Chemical equation	$Li_2O + CH_3COOH \rightarrow CH_3COOLi + H_2O$

Reaction with a metal hydroxide

Table 7.5

General equation	**metal hydroxide + carboxylic acid → salt + water**
Word equation	potassium hydroxide + methanoic acid → potassium methanoate + water
Chemical equation	$KOH + HCOOH \rightarrow HCOOK + H_2O$

Reaction with a metal carbonate

Table 7.6

General equation	**metal carbonate + carboxylic acid → salt + water + carbon dioxide**
Word equation	calcium carbonate + propanoic acid → calcium propanoate + water + carbon dioxide
Chemical equation	$CaCO_3 + CH_3CH_2COOH \rightarrow (CH_3CH_2COO)_2Ca + H_2O + CO_2$

Examples

1 **Name the salts formed from the following reactions:**
   a) **sodium hydroxide + propanoic acid**
   b) **lithium carbonate + methanoic acid**
   c) **potassium oxide + butanoic acid**
   d) **magnesium carbonate + hexanoic acid**
   e) **calcium oxide + octanoic acid**

Solution
   a) sodium propanoate
   b) lithium methanoate
   c) potassium butanoate
   d) magnesium hexanoate
   e) calcium octanoate

2 **Complete Table 7.7 below to show the reactants or products required.**

Table 7.7

Reactants	Products
lithium hydroxide + butanoic acid	a)
b)	magnesium pentanoate + water + carbon dioxide
calcium oxide + methanoic acid	c)
d)	sodium butanoate + water

Solution
   a) lithium butanoate + water
   b) magnesium carbonate + pentanoic acid
   c) calcium methanoate + water
   d) sodium oxide + butanoic acid OR sodium hydroxide + butanoic acid

Physical properties of the carboxylic acids

The carboxyl group is polar as shown in Figure 7.3.

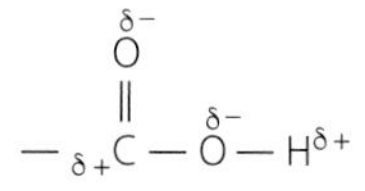

$$-\overset{\delta+}{C}-\overset{\delta-}{\underset{\displaystyle\|\atop\overset{\delta-}{O}}{}}O-H^{\delta+}$$

Figure 7.3 The carboxyl group

This allows carboxylic acids to form very effective hydrogen bonds to neighbouring carboxylic acid molecules. Figure 7.4 illustrates the hydrogen bonding between two molecules of ethanoic acid.

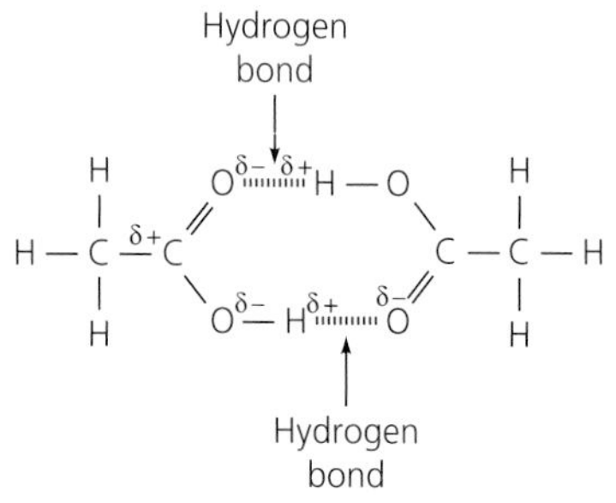

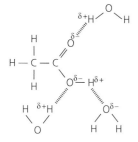

Figure 7.4 Hydrogen bonding between two molecules of ethanoic acid

Remember

Carboxylic acids are polar molecules and can form hydrogen bonds.

Thus, carboxylic acids have higher melting and boiling points than other compounds with a similar molecular mass since these strong attractive forces must be overcome in order to melt or boil the compound.

The fact that carboxylic acids can hydrogen bond allows them to dissolve in water and other polar solvents such as alcohols. Figure 7.5 illustrates how ethanoic acid dissolves in water by hydrogen bonding to water molecules.

Figure 7.5 Ethanoic acid hydrogen bonding to water molecules

Shorter chain carboxylic acids (such as methanoic and ethanoic acid) are much more soluble in water than longer chain carboxylic acids (such as octanoic acid).

Key points

* Carboxylic acids are compounds containing the carboxyl group (–COOH) and can be named using systematic naming rules.

* Carboxylic acids can react with bases (such as metal oxides, metal hydroxides and metal carbonates) to form a salt and water. With metal carbonates, carbon dioxide is also produced.

* The salts formed from these reactions can be named by changing the '-oic' ending of the acid to '-oate.'

* Carboxylic acids are polar compounds and will form hydrogen bonds. This accounts for their high melting and boiling points and their solubility in polar solvents, such as water.

Study questions

1 Draw structural formula for the following carboxylic acids:
   a) methanoic acid
   b) propanoic acid
   c) pentanoic acid
   d) 2-methylbutanoic acid
   e) 3,4-dimethylhexanoic acid
   f) 2,3,3-trimethylpentanoic acid

2 State the systematic names of the compounds shown below:

a)

```
    H   H   H   H        O
    |   |   |   |       //
H — C — C — C — C — C
    |   |   |   |       \
    H   H  CH₃  H         OH
```

b)

```
    H  CH₃  H         O
    |   |   |        //
H — C — C — C — C
    |   |   |        \
    H   H  CH₃         OH
```

c)

```
    H  CH₃  H  CH₃ H        O
    |   |   |   |   |      //
H — C — C — C — C — C — C
    |   |   |   |   |      \
    H   H   H  CH₃  H        OH
```

d)

```
         H
         |
     H — C — H
         |    H   H        O
         |    |   |       //
H — C — C — C — C
     |       |   |        \
     H      CH₃  H          OH
```

Figure 7.6

3 State the names of the products formed from the following reactions:
   a) magnesium oxide + butanoic acid →
   b) lithium oxide + ethanoic acid →
   c) sodium hydroxide + butanoic acid →
   d) magnesium carbonate + methanoic acid →

4 Name the following salts:
   a) HCOOLi
   b) CH_3COONa
   c) $(CH_3COO)_2Ca$
   d) $(CH_3CH_2COO)_2Mg$

5 Draw a full structural formula for hexanoic acid. By showing the partial charges on the molecule, explain how hexanoic acid can dissolve in water.

6 Why does ethanoic acid have a higher boiling point than ethanol?

7 Ammonium hydroxide ($NH_4^+OH^-$) can react with ethanoic acid to form a salt. Name this salt and draw a structural formula for this salt.

Chapter 8
Esters, fats and oils

What you should know

★ An ester is a molecule containing an ester link: –COO–.
★ Esters are formed by a condensation reaction between an alcohol and a carboxylic acid.
★ In a condensation reaction, two molecules are joined together with the elimination of a small molecule. When an ester link is formed by the reaction between a hydroxyl group and a carboxyl group, the small molecule eliminated is water.
★ Esters can be named, and molecular or structural formulae drawn, given the names of their parent alcohol and carboxylic acid or structural formulae of the ester.
★ Esters are used as flavourings and fragrances as many have pleasant, fruity smells. Esters are also used as solvents for non-polar compounds that do not dissolve in water.
★ In a hydrolysis reaction, a molecule reacts with water to break down into smaller molecules. Esters can be hydrolysed to produce an alcohol and a carboxylic acid.
★ The products of the hydrolysis of an ester can be named given the name or molecular formula or structural formula of the ester.
★ Edible fats and edible oils are esters formed from the condensation of glycerol (propane-1,2,3-triol) and three carboxylic acid molecules. The carboxylic acids are known as 'fatty acids' and can be saturated or unsaturated straight-chain carboxylic acids, usually with long chains of carbon atoms.
★ Edible oils have lower melting points than edible fats.
★ Double bonds in fatty acid chains prevent oil molecules from packing closely together, so the greater the number of double bonds present, the weaker the van der Waals' forces of attraction. The greater the degree of unsaturation, the lower the melting point.
★ Unsaturated compounds quickly decolourise bromine solution.
★ The bromine molecules add across the carbon–carbon double bonds in an addition reaction. The greater the number of double bonds present in a substance, the more bromine solution can be decolourised.
★ Fats and oils are
    ★ a concentrated source of energy
    ★ essential for the transport and storage of fat-soluble vitamins in the body.

Making an ester

Some of the most memorable smells come from compounds known as
esters. Esters are formed by reacting an alcohol with a carboxylic acid.
This reaction is known as a **condensation reaction** since the alcohol and
carboxylic acid join together by eliminating water.

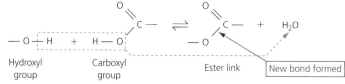

Figure 8.1 Forming an ester from ethanol and ethanoic acid

Condensation reactions

Condensation reactions can involve the elimination of small molecules other than water. For example, a common method for forming an ester involves reacting an alcohol with a compound called an acid chloride as shown below. In this case, the small molecule eliminated is hydrogen chloride (HCl).

$$
\begin{array}{c}
\mathrm{H} \quad \mathrm{O} \\
| \quad \| \\
\mathrm{H}-\mathrm{C}-\mathrm{C}-\mathrm{Cl} \\
| \\
\mathrm{H}
\end{array}
+ \mathrm{H}-\mathrm{O}-\mathrm{CH_3} \quad \rightleftharpoons \quad
\begin{array}{c}
\mathrm{H} \quad \mathrm{O} \\
| \quad \| \\
\mathrm{H}-\mathrm{C}-\mathrm{C}-\mathrm{O}-\mathrm{CH_3} \\
| \\
\mathrm{H}
\end{array}
+ \mathrm{H}-\mathrm{Cl}
$$

Figure 8.2 Condensation reactions

Naming esters

Table 8.1 shows the names of some common esters, their structures and the alcohol and carboxylic acid from which they were made.

Table 8.1 Naming esters

Alcohol	Carboxylic acid	Ester formed
Methanol $$\begin{array}{c} \mathrm{H} \\ \| \\ \mathrm{H}-\mathrm{C}-\mathrm{O}-\mathrm{H} \\ \| \\ \mathrm{H} \end{array}$$	Ethanoic acid	Methyl ethanoate
Ethanol	Propanoic acid	Ethyl propanoate
Ethanol	Butanoic acid	Ethyl butanoate

The first part of the ester name, which ends in '-yl', comes from the alcohol and the second part, which ends in '-oate', comes from the acid. Two examples are shown in Figures 8.3 and 8.4.

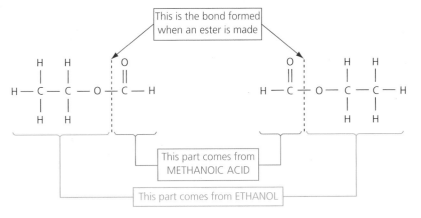

This is the bond formed when an ester is made

This part comes from METHANOIC ACID

This part comes from ETHANOL

Hence, this ester is called ETHYL METHANOATE

Figure 8.3 Naming an ester using full **structural formulae**

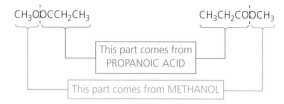

Hence, this ester is called METHYL PROPANOATE

Figure 8.4 Naming an ester using shortened structural formulae

Uses of esters

As many esters have a pleasant, fruity smell, they are often used in scented products such as deodorants and perfumes. Pleasant tasting esters are also used as flavourings. For example, propyl pentanoate is used to give a pineapple flavour to foods.

Esters are widely used as solvents for non-polar compounds. They are able to dissolve a wide variety of compounds and have a low boiling point, allowing them to evaporate easily. For example, ethyl ethanoate is used to dissolve coloured compounds used in nail varnish. Once applied to the nail, the ethyl ethanoate evaporates to leave the coloured compound on the nail. Similarly, esters are used as solvents for paints.

Figure 8.5 Nail varnish contains the ester ethyl ethanoate as the solvent.

Example

Name the following esters:

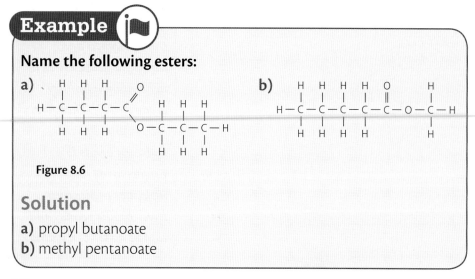

Figure 8.6

Solution

a) propyl butanoate
b) methyl pentanoate

Hydrolysis of esters

When an ester reacts with water, the ester breaks down to form an alcohol and carboxylic acid. This reaction is known as **ester hydrolysis**. For example, the ester methyl ethanoate will form methanol and ethanoic acid when it is hydrolysed. Table 8.2 shows the products formed when the ester is hydrolysed.

Table 8.2 Ester hydrolysis produces an alcohol and a carboxylic acid.

Ester	Alcohol formed	Carboxylic acid formed
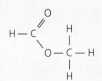		

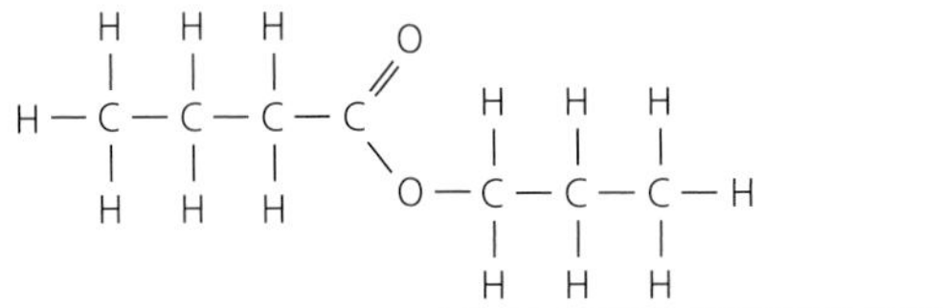

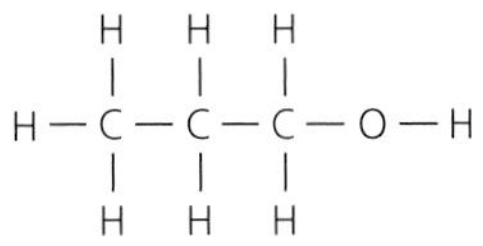

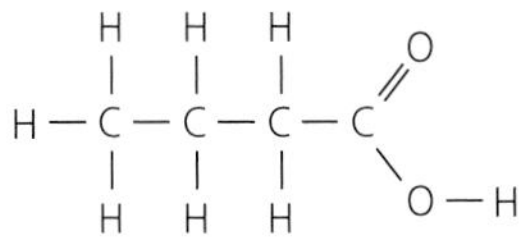

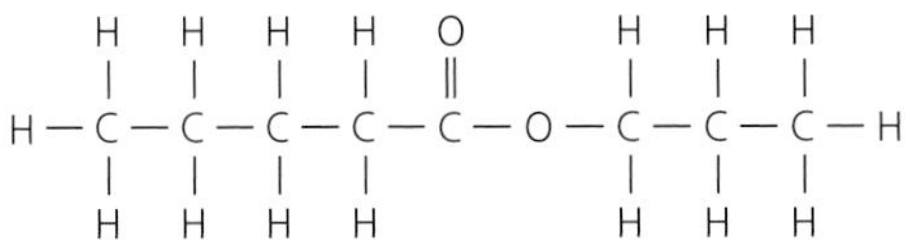

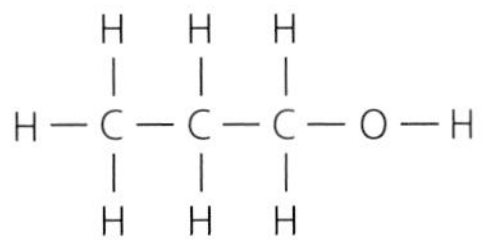

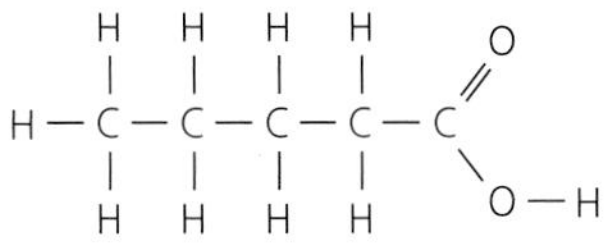

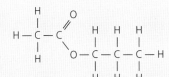

Examples

1 **State the names of the compounds formed when the following esters are hydrolysed:**
   a) **ethyl propanoate**
   b) **methyl hexanoate**
   c) **propyl octanoate**
2 **Name the esters shown below and draw full structural formulae for the compounds formed when the esters are hydrolysed.**
   a)
   b)

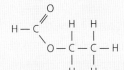

   c)

Figure 8.7

Solutions

1 **a)** ethanol and propanoic acid
   b) methanol and hexanoic acid
   c) propan-1-ol and octanoic acid

> *Remember*
>
> Forming an ester is a condensation reaction.
> Breaking up an ester is a hydrolysis reaction.

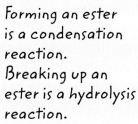

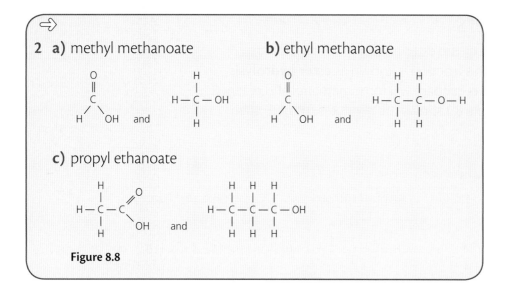

2 a) methyl methanoate   **b)** ethyl methanoate

c) propyl ethanoate

Figure 8.8

Fats and oils

Edible **fats** and edible **oils** are esters found in a variety of foods such as milk, butter, meat, fish, nuts, etc. They are a special type of ester known as a tri-ester as they contain three ester links, as shown in Figure 8.10.

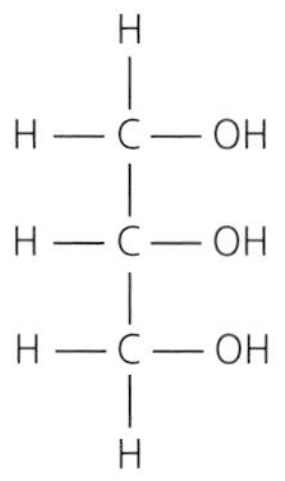

Propane-1,2,3-triol or glycerol

Figure 8.9 Glycerol is an alcohol with three –OH groups.

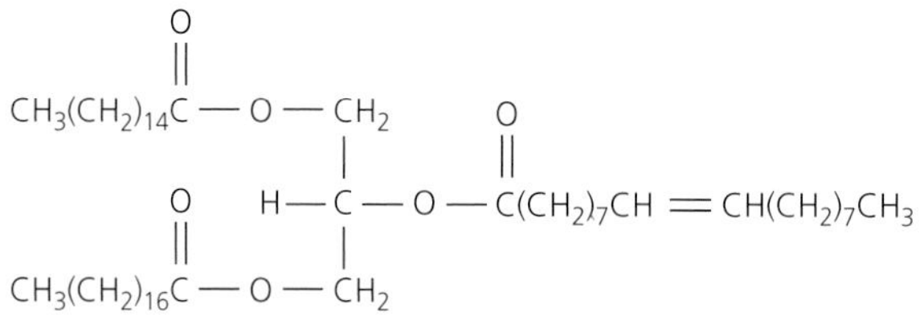

$CH_3(CH_2)_{14}C$
$CH_3(CH_2)_{16}C$

$H-C-O-C(CH_2)_7CH = CH(CH_2)_7CH_3$

Figure 8.10 An example of an oil molecule. Note the three ester links.

This arises from the fact that all edible fats and oils are formed from the alcohol glycerol (propane-1,2,3-triol), which contains three hydroxyl groups. This allows glycerol to react with three carboxylic acid molecules. The type of carboxylic acid molecules that react with glycerol to form a fat or oil are known as **fatty acids**. These are carboxylic acids that are found in nature and contain many carbon atoms (e.g. 20 per molecule).

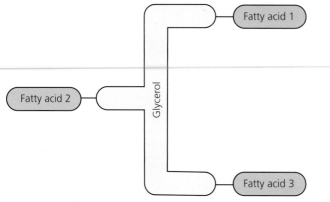

Figure 8.11 The structure of a fat or oil

At room temperature fats are solid whereas oils are liquids. The structure of the fatty acids which join glycerol affects whether the **triglyceride** formed is a fat or an oil.

Edible oils have lower melting points than edible fats. This arises from their structures. In an oil, the fatty acids are unsaturated. In a fat, the fatty acids are saturated. This fact means that the shape of an oil molecule is different from the shape of a fat molecule, which is illustrated in Figure 8.12.

Hints & tips

Saturated fats are Solid.

The efficient packing of fat molecules allows for the intermolecular forces of attraction (LDF) to hold the fat molecules together. These attractive forces are greater between fat molecules than between oil molecules because of this efficient packing. When the fat is heated, it is these London dispersion forces between the fat molecules that are broken, allowing the fat molecules to separate.

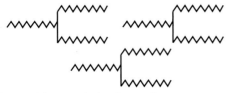

Figure 8.12 Fat molecules are able to pack closely together.

In oils, the double bonds cause the molecules to have an uneven, distorted structure, which makes it difficult for them to pack closely together, as shown in Figure 8.13. This results in fewer attractions (LDF) between oil molecules. It therefore takes less energy to break apart the oil molecules which means that oils have a lower melting point than fats.

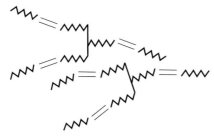

Figure 8.13 Oil molecules cannot pack closely together.

As it is the double bonds in the fatty acid chains that prevent oil molecules from packing closely together, the greater the number of double bonds present, the weaker the forces of attraction. So, oils which have a greater number of double bonds will have a lower melting point.

One way to test the 'degree of unsaturation' (i.e. the number of double bonds in an oil) is to react the oil with bromine solution since oils can decolourise bromine solution due to the presence of their double bonds. A simple test can be carried out to determine what volume of bromine solution is required to react with a fixed quantity of oil. Oils with a higher degree of unsaturation will require a higher volume of bromine solution than oils with a lower degree of unsaturation.

The reaction of bromine solution with an oil is an example of an addition reaction.

Fats and oils in the diet

As well as being a concentrated source of energy, fats and oils are an essential component of our diet as their non-polar structure allows them to dissolve and transport a number of essential vitamins, such as vitamins A and D.

Key points

* Esters are formed by the condensation reaction of an alcohol and a carboxylic acid.
* In a condensation reaction, two molecules come together to form a larger molecule and a small molecule, such as water.
* The name of an ester comes from the alcohol and carboxylic acid from which it is formed. For example, reacting ethanol and methanoic acid produces the ester ethyl methanoate.
* Breaking down an ester is an example of a hydrolysis reaction.
* Esters often have fruity smells and tastes and are used as flavourings and scents.
* Esters are insoluble in water and can be used as solvents for non-polar compounds.
* Edible fats and edible oils are examples of esters formed from the reaction of glycerol (propane-1,2,3-triol) with three fatty acids.
* Fat molecules are mainly saturated and can pack closely together making them solid at room temperature as there are lots of intermolecular forces of attraction (LDF) holding the molecules together.
* Oil molecules have a large number of double bonds and cannot pack closely together, making them liquids at room temperature as there are fewer intermolecular forces of attraction between the molecules.
* The degree of unsaturation in an oil can be measured by reacting an oil with bromine solution. The more bromine solution required to react, the more double bonds present.
* Oils with a high degree of unsaturation will have lower melting points than those with a lower degree of unsaturation.
* Edible fats and edible oils are essential in the diet for dissolving and transporting fat soluble vitamins. They are also a source of energy.

Study questions

1 Draw full chemical structures for the following esters:
   a) methyl butanoate
   b) ethyl propanoate
   c) methyl ethanoate

2 Draw the structures of
   a) the hydroxyl group
   b) the carboxyl group
   c) the ester link.

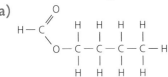

3 Name the following esters:

a)

H—C(=O)O—C(H)(H)—C(H)(H)—C(H)(H)—C(H)(H)—H

Figure 8.14

b)

H—C(H)(H)—C(H)(H)—C(H)(H)—C(H)(H)—C(=O)—O—C(H)(H)—C(H)(H)—C(H)(H)—H

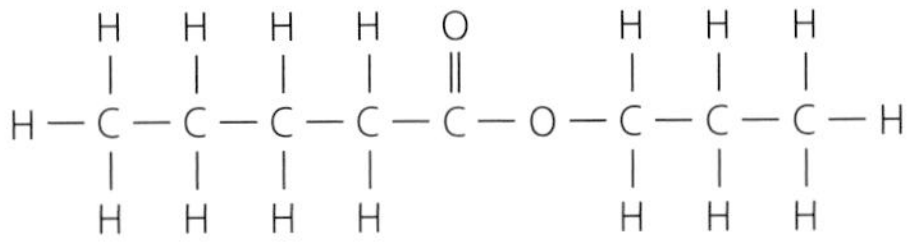

Figure 8.15

4 Name the products formed when the esters in Question 1 are hydrolysed.

5 The ester shown in Question 3(a) was hydrolysed using KOH(aq). Draw a structural formula for the salt of the carboxylic acid that would be formed.

6 Name the type of reaction:
  a) when an alcohol reacts with a carboxylic acid to form an ester
  b) when an ester reacts with water to form an alcohol and a carboxylic acid.

7 Which of the following cannot be applied to glycerol?
  A It is known as propane-1,2,3-triol.
  B It can react with three carboxyl groups.
  C It has a low viscosity.
  D It can hydrogen bond.

8 Which of the following statements cannot be applied to fats and oils?
  A They are esters.
  B They can hydrolyse to produce fatty acids.
  C They are likely to be soluble in hexane.
  D They can react with glycerol to form esters.

9 Vitamins can be classed as being 'fat soluble' or 'water soluble'. The structures of vitamin C, a water-soluble vitamin, and vitamin A, a fat-soluble vitamin, are shown in Figures 8.16 and 8.17.

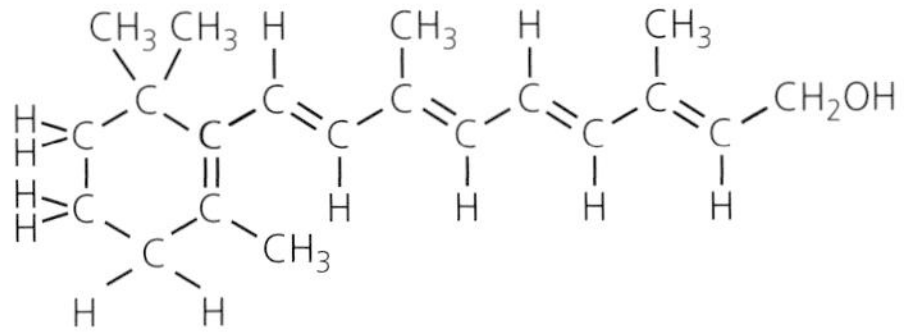

Figure 8.16 Vitamin A, fat soluble

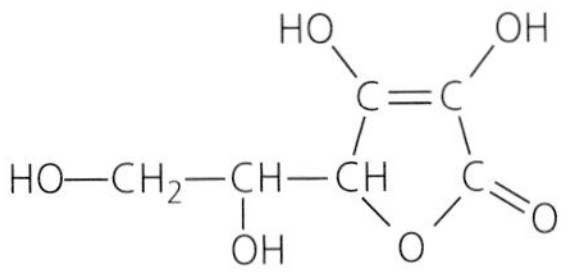

Figure 8.17 Vitamin C, water soluble

  a) In terms of structure, why is vitamin C a water-soluble vitamin?
  b) In terms of structure, why is vitamin A a fat-soluble vitamin?

10 Explain why oil molecules have lower melting points than fat molecules.

11 a) Explain why the molecule shown as Figure 8.18 is known as a monoglyceride.

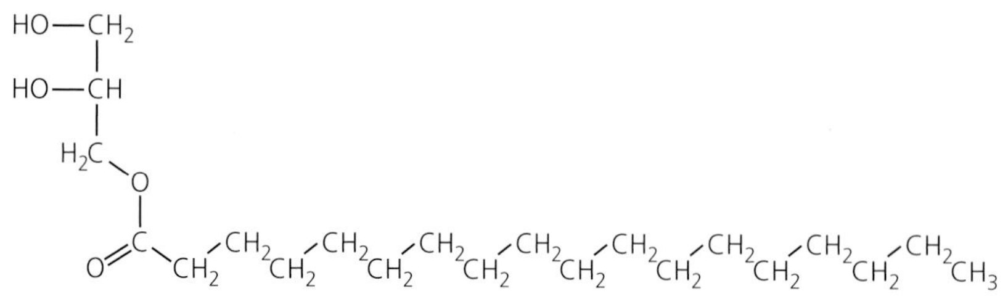

$HO-CH_2$
$HO-CH$
H_2C
O
C
O CH_2 CH_2 CH_2 CH_2 CH_2 CH_2 CH_2 CH_2
CH_2 CH_2 CH_2 CH_2 CH_2 CH_2 CH_2 CH_3

Figure 8.18

b) Which part of the monoglyceride could dissolve in a non-polar solvent such as hexane?

12 A student tested two fatty acids with bromine solution to investigate which fatty acid had the highest degree of unsaturation.

The formulae for the fatty acids are shown below:

Fatty acid A: $C_{17}H_{31}COOH$

Fatty acid B: $C_{17}H_{35}COOH$

a) Name the type of reaction the fatty acid will take part in if it reacts with bromine solution.

b) State which fatty acid would have the highest degree of unsaturation. Explain your answer.

Chapter 9

Soaps, detergents and emulsions

What you should know

★ Soaps are produced by the alkaline hydrolysis of edible fats and edible oils. Hydrolysis produces three fatty acid molecules and one glycerol molecule. The fatty acid molecules are neutralised by the alkali, forming water-soluble, ionic salts, called soaps.

★ Soaps can be used to remove non-polar substances such as oil and grease. Soap ions have long non-polar tails, readily soluble in non-polar compounds (hydrophobic), and ionic heads that are water soluble (hydrophilic). The hydrophobic tails dissolve in the oil or grease. The negatively-charged hydrophilic heads remain in the surrounding water.

★ Agitation causes ball-like structures to form. The negatively-charged ball-like structures repel each other and the oil or grease is kept suspended in the water.

★ 'Hard water' is a term used to describe water containing high levels of dissolved metal ions.

★ When soap is used in hard water, scum, an insoluble precipitate, is formed.

★ Soapless detergents are substances with non-polar hydrophobic tails and ionic hydrophilic heads. These remove oil and grease in the same way as soap. Soapless detergents do not form scum with hard water.

★ An emulsifier can be used to prevent non-polar and polar liquids separating into layers. An emulsion contains small droplets of one liquid dispersed in another liquid.

★ Emulsifiers for use in food can be made by reacting edible oils with glycerol. In the molecules formed, only one or two fatty acid groups are linked to each glycerol backbone.

★ The hydroxyl groups present in the emulsifier are hydrophilic, while the fatty acid chains are hydrophobic. The hydrophobic fatty acid chains dissolve in oil, while the hydrophilic hydroxyl groups dissolve in water, forming a stable emulsion.

Edible fats and edible oils are examples of esters. When they hydrolyse, they produce glycerol and carboxylic acids known as fatty acids. If the hydrolysis is carried out using an alkali such as NaOH(aq), the fatty acids immediately react to form salts. This is illustrated by the reaction shown in Figure 9.1. These salts are what we commonly refer to as soaps.

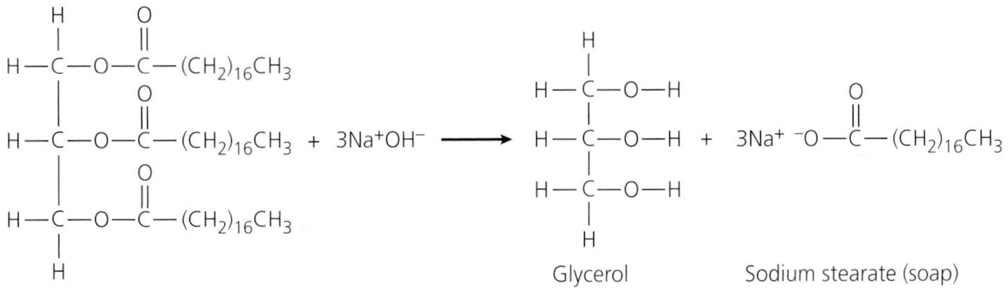

Figure 9.1 Alkaline hydrolysis of a fat produces glycerol and soap.

The structure and cleansing mechanism of soap

Edible fats and edible oils are the main compounds found in greasy stains from foods. Since fats and oils are non-polar compounds, they will not readily dissolve in water. Instead, a soap is required to help dissolve the greasy stain and remove it from clothes or skin.

As shown in Figure 9.1, a soap is the salt of a carboxylic acid. Since the carboxylic acids found in fats and oils have long hydrocarbon chains, soap molecules also contain long, non-polar hydrocarbon chains. In addition, soaps contain a carboxylate 'head' formed from the reaction of the carboxyl group (–COOH) with the base. Overall, soaps are said to have a polar head and a non-polar tail. This is represented in Figures 9.2 and 9.3.

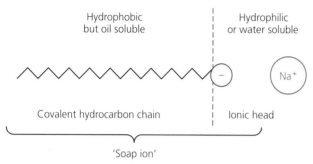

Figure 9.2 The structure of a soap

$$CH_3(CH_2)_{14}COO^-$$

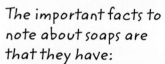

Hydrophobic tail   Hydrophilic head

Figure 9.3 A soap ion

The hydrocarbon tail can bond easily to greasy stains on clothes or skin. This part of the soap molecule does not dissolve in water; it is **hydrophobic**. The ionic head does dissolve in water; it is **hydrophilic**. As shown in Figure 9.4, the soap molecules can dissolve in the grease causing it to be covered in negative charge. Agitation (mixing the soap, water and grease) causes ball-like structures to form, known as globules. The negatively-charged globules repel each other and the oil or grease is kept suspended in the water. In other words, the oil/grease is no longer a separate layer. It is broken up into globules, which are now attracted to water allowing them to be washed away.

> **Remember**
>
> The important facts to note about soaps are that they have:
>
> 1 an ionic head which is water soluble
> 2 a hydrocarbon tail which is soluble in oil and grease.

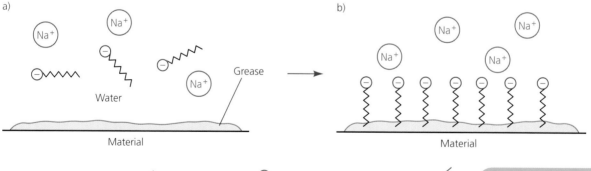

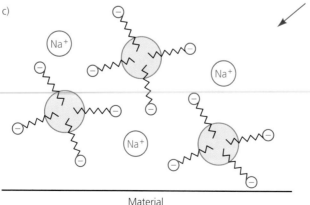

Figure 9.4 Greasy material being washed by soapy water. Note that the sodium ions are dissolved in the water.

> **Hints & tips**
>
> Soaps work because they have a water-soluble head and an oil-soluble tail. You should be able to describe the structure of a typical soap molecule and understand why the head and tail have these properties.

Detergents

Detergents are compounds with a soap-like structure which allows them to remove greasy stains. They have a non-polar tail, which is oil soluble, and a polar head, which is water soluble. An example of a detergent molecule is dodecylbenzenesulfonate: $CH_3(CH_2)_{11}C_6H_4SO_3^-$.

The main difference in structure between a soap and a detergent is the head. The head part of a detergent will always be soluble in water. This can be achieved by using heads which are negative ions (such as the sulfonate ion), positive ions or have an overall polar structure. The detergent head will never, however, be formed from a carboxylate ion.

The water supply in many regions in the UK is known as 'hard water' as it is rich in dissolved calcium and magnesium ions. When this water is mixed with soap, instead of forming a soapy lather, a precipitate is formed:

$$CH_3(CH_2)_{14}COO^-(aq) + Ca^{2+}(aq) \rightarrow (CH_3(CH_2)_{14}COO)_2Ca(s)$$

This precipitate reduces the cleansing action of the soap and builds up to leave a 'scum' around baths, sinks and shower heads. Detergents do not form precipitates with calcium or magnesium ions as they do not contain the carboxylate ion. Hence, detergents are very useful in hard-water areas as an alternative to soap.

Emulsions and emulsifiers

An **emulsion** is a liquid which contains small droplets of another liquid, such as soapy grease particles mixed in water. The grease and water would normally separate into two separate layers, but the soap acts as an **emulsifier** as it helps to bring the two substances together. Emulsions in food are very common. Milk is mostly water but it also contains lots of fats. If you look at a glass of milk, you do not see the separate layers as milk contains emulsifiers which help the fatty parts of milk remain dispersed throughout the water.

Mayonnaise is an example of an emulsion formed by mixing olive oil and vinegar. Normally, the olive oil and vinegar would not mix and two separate layers would form. However, mayonnaise also comprises egg yolk which contains the compound lecithin. Lecithin acts as an emulsifier as it contains a charged part (which is water soluble) and a large hydrocarbon tail (which is non-polar). The vinegar is attracted to the charged part of lecithin while the olive oil is attracted to the hydrocarbon part of lecithin.

Many of the emulsifiers used in the food industry are formed from edible oils. Instead of containing three ester groups, they contain one or two and are known as monoglycerides or diglycerides, respectively. The main point is that they will have at least one hydroxyl group (−OH). This is the polar part of the molecule, which can attract water while the long hydrocarbon chains will attract non-polar molecules. For example, in ice cream, to stop the water separating from the fatty molecules, a monoglyceride emulsifier is often used as the hydroxyl groups dissolve in water and the

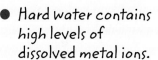

Figure 9.5 Milk and mayonnaise are examples of emulsions.

long chain hydrocarbon tails can dissolve in the fatty parts of the ice cream. Without an emulsifier, ice cream would form two separate layers.

R_1 and R_2 are long, hydrophobic, hydrocarbon chains.

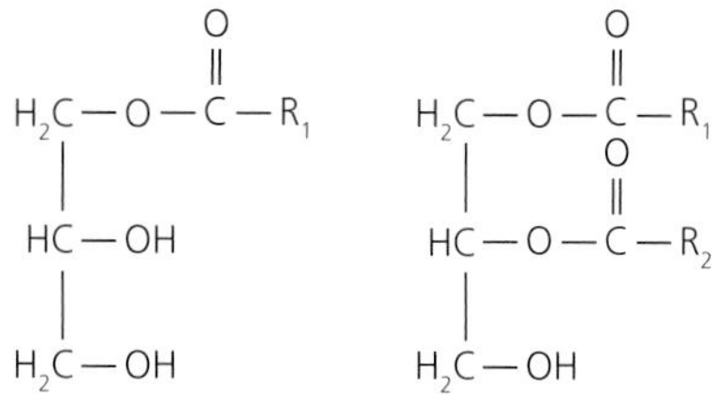

Monoglyceride   Diglyceride

Figure 9.6 Monoglycerides and diglycerides are emulsifiers formed from edible oils.

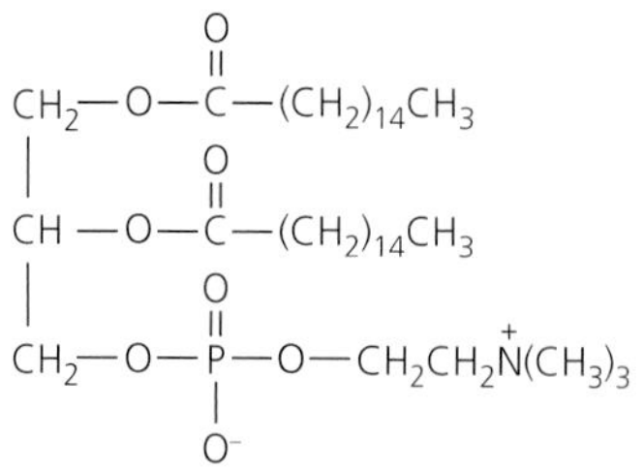

Figure 9.7 Lecithin is found in egg yolk and acts as an emulsifier.

Key points !

* Soaps are fatty acid salts formed from the alkaline hydrolysis of fats or oils.
* Soap molecules contain a non-polar tail and a polar head.
* Soaps can dissolve greasy/oily stains as the non-polar tail will dissolve in the grease, coating the outside of the stain with repulsive negative charges from the polar heads. This causes the stain to break down into water-soluble globules.
* Detergents are soap-like molecules which will not form scummy precipitates with hard water as they have different polar heads from soaps.
* Emulsifiers are used to keep water-soluble and oil-soluble compounds together. They are widely used in the food industry and can be recognised as they have a non-polar part and a polar part.

Study questions ?

1   Which of the following compounds could act as a soap?
    A   stearic acid   C   sodium stearate
    B   glycerol   D   propyl stearate

2   Detergents are useful replacements for soaps as they
    A   do not dissolve in water   C   do not dissolve in oil
    B   do not form precipitates with water   D   do not contain glycerol.

3   The following compound is an example of an emulsifier.

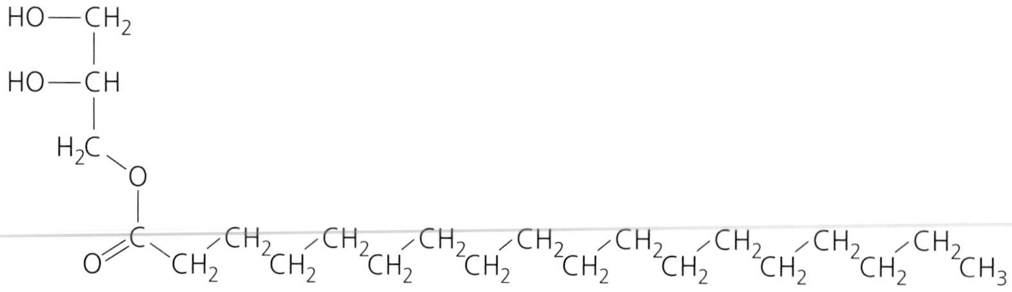

Figure 9.8

   a) Draw a structural formula for the alcohol formed when this compound is hydrolysed.
   b) Explain how this compound can act as an emulsifier.

4 A representation of the detergent molecule sodium dodecyl sulfate is shown in Figure 9.9.

Figure 9.9

a) Part of the detergent molecule is circled. Suggest what this represents.

b) Explain how a detergent molecule can dissolve an oily stain.

Chapter 10
Proteins

What you should know

★ Proteins are the major structural materials of animal tissue and are also involved in the maintenance and regulation of life processes. Enzymes are proteins which act as biological catalysts.

★ Amino acids, the building blocks from which proteins are formed, are relatively small molecules which all contain an amino group, —NH_2, and a carboxyl group, —COOH.

★ Proteins are made of many amino acid molecules linked together by condensation reactions. In these reactions, the amino group of one amino acid and the carboxyl group of another amino acid join, with the elimination of water.

★ The link which forms between two amino acids is known as a peptide link, —CONH—, or also as an amide link.

★ Proteins which fulfil different roles in the body are formed by linking together differing sequences of amino acids.

★ The body cannot make all of the amino acids required for protein synthesis and certain amino acids, known as essential amino acids, must be acquired from the diet.

★ During digestion, enzyme hydrolysis of protein produces amino acids.

★ The structural formulae of amino acids obtained from the hydrolysis of a protein can be drawn given the structure of a section of the protein.

★ The structural formula of a section of protein can be drawn given the structural formulae of the amino acids from which it is formed.

★ Within proteins, the long-chain molecules form spirals, sheets or other complex shapes.

★ The chains are held in these forms by intermolecular bonding between the side chains of the constituent amino acids. When proteins are heated, these intermolecular bonds are broken, allowing the proteins to change shape (denature). The denaturing of proteins in foods causes the texture to change when it is cooked.

Protein structure and function

Proteins are nitrogen-containing compounds that are found in all living things. As can be seen from Table 10.1, proteins are used as structural materials in animals (e.g. they form skin, bone, teeth and hair) and are also involved in the maintenance and regulation of life processes (e.g. they are used to build **hormones**).

Table 10.1 Examples of proteins and their functions

Name of protein	Where found	Function
Collagen	Tendons, muscle and bone	Gives structural support
Keratin	Hair, skin and nails	Gives structural support
Myosin	Muscles	Helps muscles to contract
Insulin	Pancreas	Hormone which helps to control blood glucose levels
Haemoglobin	Red blood cells	Transports oxygen around the body
Immunoglobulins	Blood, tears, saliva, skin	Fight infection
Amylase	Saliva and pancreas	An enzyme which breaks down starch

Enzymes – sometimes described as biological catalysts – are examples of proteins. The specific shape of a protein allows it to catalyse a specific reaction.

All proteins are made up of smaller molecules known as **amino acids**. Amino acids contain an **amino** group ($-NH_2$) and a carboxyl group ($-COOH$). Figure 10.1 shows the typical structure of an amino acid. Figure 10.2 shows the structures of six common amino acids.

$$H_2N - \overset{\overset{\displaystyle H}{|}}{\underset{\underset{\displaystyle H}{|}}{C}} - COOH$$

Glycine (gly)

$$H_2N - \overset{\overset{\displaystyle H}{|}}{\underset{\underset{\displaystyle CH_3}{|}}{C}} - COOH$$

Alanine (ala)

$$H_2N - \overset{\overset{\displaystyle H}{|}}{\underset{\underset{\displaystyle CH_2-SH}{|}}{C}} - COOH$$

Cysteine (cys)

$$H_2N - \overset{\overset{\displaystyle H}{|}}{\underset{\underset{\displaystyle CH_2}{|}}{C}} - COOH$$

Phenylalanine (phe)

$$H_2N - \overset{\overset{\displaystyle H}{|}}{\underset{\underset{\displaystyle CH_2OH}{|}}{C}} - COOH$$

Serine (ser)

$$H_2N - \overset{\overset{\displaystyle H}{|}}{\underset{\underset{\displaystyle CH_2COOH}{|}}{C}} - COOH$$

Glutamic acid (glu)

R represents a variable organic group (or hydrogen)

Amino group   Carboxyl group

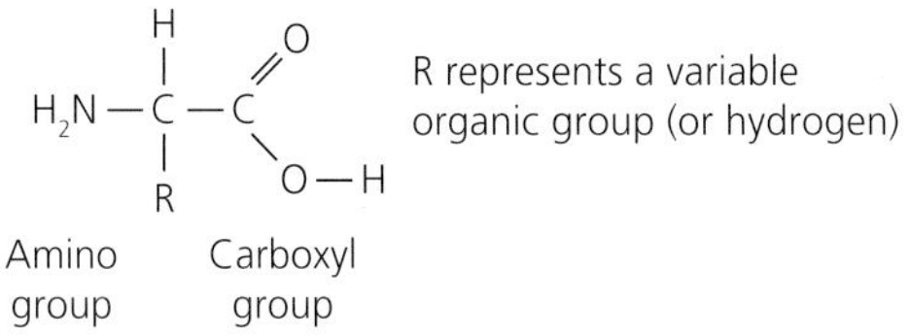

Figure 10.1 Amino acid structure

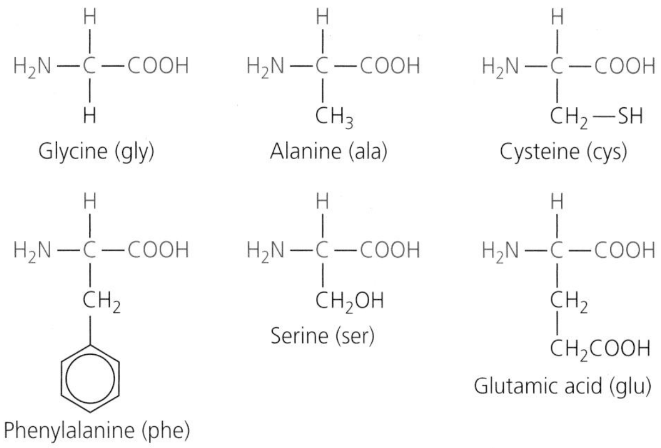

Figure 10.2 The structures of six common amino acids

Proteins are normally massive long chain molecules, made up of many thousands of amino acids joined together. It is the sequence of amino acids that gives rise to different proteins. For example, joining the amino acids glycine to alanine to serine (gly-ala-ser) will produce a different molecule with different properties if the amino acids are joined in a different order, e.g. alanine to serine and then glycine (ala-ser-gly).

Making a protein

Several amino acid molecules react together to form a protein molecule. This reaction involves the amino group reacting with the carboxyl group to form an **amide link**. Water is formed when this reaction takes place. As amino acids are joined by elimination of water, the reaction of amino acids to form a protein is an example of a condensation reaction. This is illustrated in Figure 10.3. Note that the amide link is also known as the **peptide link**.

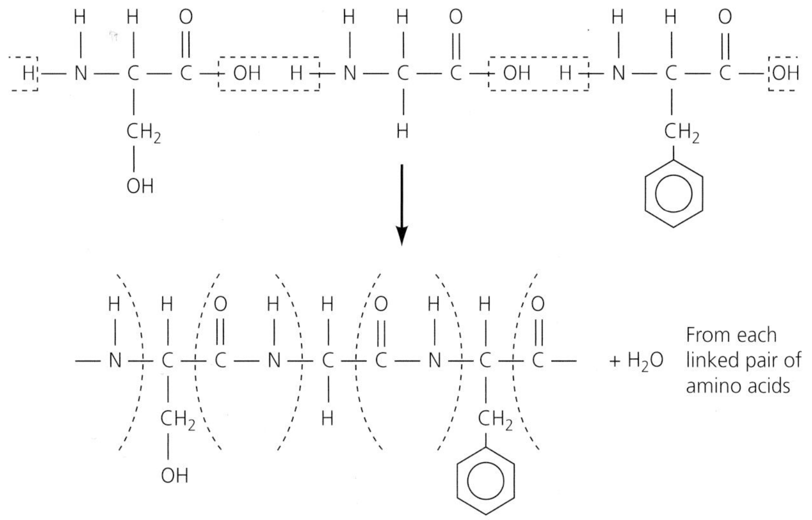

+ H_2O From each linked pair of amino acids

Figure 10.3 Condensation reaction of amino acids to form a protein. The amide links are shown in brackets.

Remember

The structure of the amide link is:

$$- CONH - \text{ or } - \overset{\overset{\displaystyle O}{\|}}{\underset{\underset{\displaystyle H}{|}}{C}} - N -$$

Figure 10.4

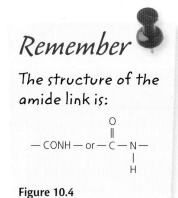

Example

A tripeptide can be formed by joining together three amino acids. Draw the tripeptide formed by linking together three molecules of glycine (see Figure 10.6 for the structure of glycine).

Solution

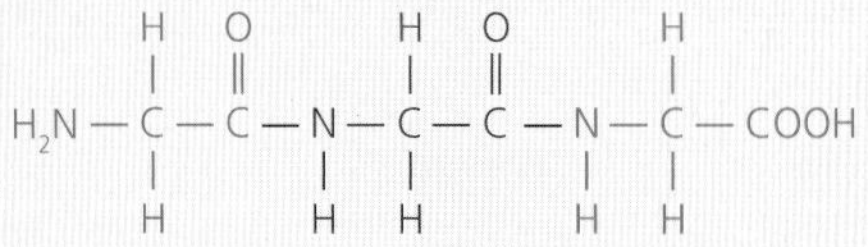

$$H_2N - C - C - N - C - C - N - C - COOH$$

Figure 10.5

Breaking a protein

The order in which amino acids join together determines the type of protein formed. The human body requires a source of amino acids in order to make the proteins needed to maintain and regulate life. Fortunately, the human body can make most of its own amino acids. There are, however, several amino acids which the body cannot make: these are known as **essential amino acids** and must be obtained from the diet.

Once eaten, proteins are broken down by the process of digestion to form the amino acids that make up the protein. This process is known as hydrolysis and is shown in Figure 10.6.

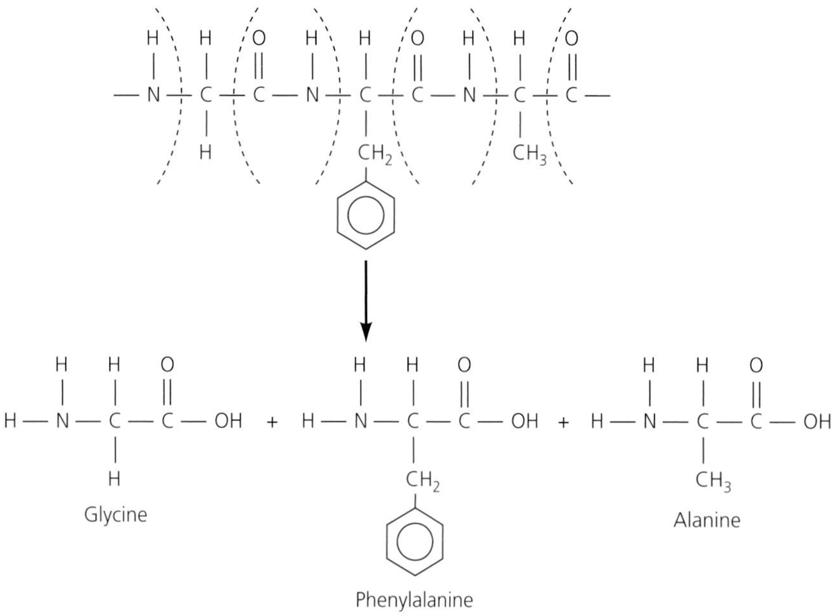

Figure 10.6 Hydrolysis of a protein to form amino acids

Glycine

Phenylalanine

Alanine

Both amino acids and proteins contain an N–H bond. This is significant as it allows hydrogen bonding to take place. Hydrogen bonding between protein molecules allows long protein strands to curl into different shapes.

When proteins are heated, the hydrogen bonds holding the chains together are broken allowing the proteins to change shape (denature). The denaturing of proteins in foods causes the texture to change when it is cooked.

> **Hints & tips**
>
> *You should be able to draw and recognise the functional groups mentioned in this chapter: amino, carboxyl and amide/peptide.*

> **Remember**
>
> Proteins can form hydrogen bonds as they contain an N–H.

The shape of an enzyme is essential for it to work. This shape is caused by the protein chains attracting each other using intermolecular forces (mainly hydrogen bonding). When heated, the intermolecular forces are broken causing the chains to unfold, i.e. the shape has changed. This causes the enzyme to stop working.

Example

Draw structures for the three amino acids formed when the molecule shown is hydrolysed.

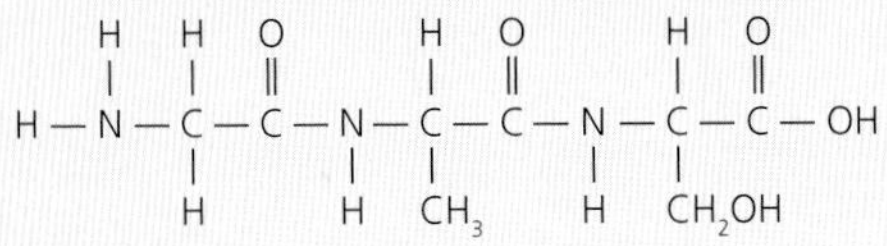

Figure 10.7

Solution

Hint: Identify the amide links then add an —OH to the C and an H to the N.

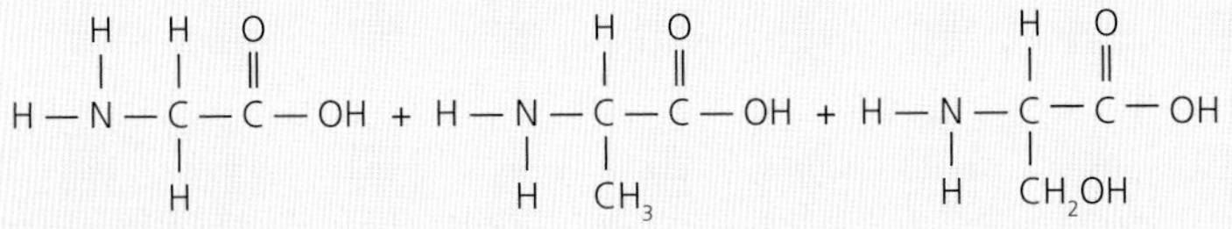

Figure 10.8

Key points !

* Amino acids are small molecules that contain an amino group ($-NH_2$) and a carboxyl group ($-COOH$).
* An amino acid can react with a neighbouring amino acid by reacting an amino group with a carboxyl group. The resultant bond formed is known as an amide (or peptide) link.
* When amino acids react together, water is eliminated. This is a condensation reaction.
* Joining amino acids together in a different sequence will produce a different protein.
* The splitting up of a protein into amino acids is known as a hydrolysis reaction.
* Essential amino acids are those amino acids required by the body which cannot be made by the body; they must be obtained from the diet.
* Heating a protein breaks the intermolecular forces (mainly hydrogen bonds) holding the long protein molecules together. This changes the shape of the structure and is known as 'denaturing'.

Study questions ?

1 Which of the following represents an amino acid?
   A CH_3NH_2
   B C_6H_5COOH
   C H_2NCH_2COOH
   D $NCCH_3COOH$

2 Which of the following statements correctly describes an essential amino acid?
   A An amino acid that can only be made by the body.
   B An amino acid that must be obtained from the diet.
   C An amino acid that performs a key function in the body.
   D An amino acid that is only found in one type of protein.

*3 A tripeptide X has the structure shown in Figure 10.9. Partial hydrolysis of X yields a mixture of dipeptides. Which of the dipeptides shown in Figure 10.10 could be produced on hydrolysing X?

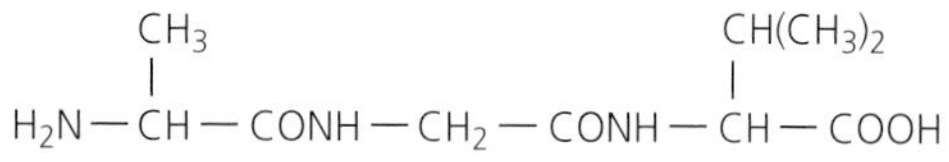

$$CH_3 \qquad\qquad\qquad CH(CH_3)_2$$
$$|\qquad\qquad\qquad\qquad\qquad |$$
$$H_2N-CH-CONH-CH_2-CONH-CH-COOH$$

Figure 10.9

A
$$\qquad\qquad\qquad\qquad CH_3$$
$$\qquad\qquad\qquad\qquad |$$
$$H_2N-CH_2-CONH-CH-COOH$$

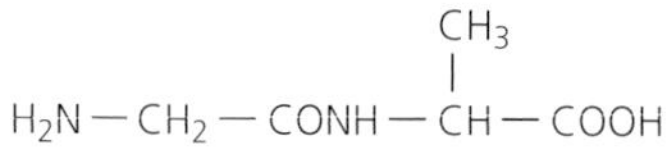

B
$$\qquad CH_3 \qquad\quad CH(CH_3)_2$$
$$\qquad | \qquad\qquad\quad |$$
$$H_2N-CH-CONH-CH-COOH$$

C
$$\qquad CH(CH_3)_2$$
$$\qquad |$$
$$H_2N-CH-CONH-CH_2-COOH$$

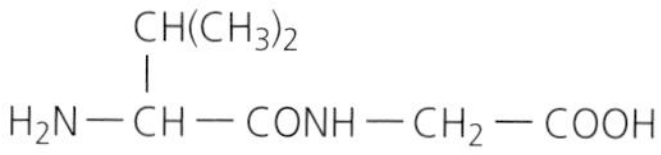

D
$$\qquad\qquad\qquad CH(CH_3)_2$$
$$\qquad\qquad\qquad |$$
$$H_2N-CH_2-CONH-CH-COOH$$

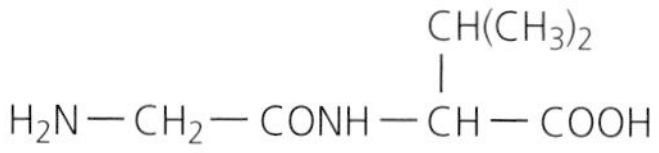

Figure 10.10

4 Which statement does not apply to amino acids?

   A They can react with bases to form salts.
   B They can react with glycerol to form fats.
   C They can react with other amino acids to form proteins.
   D They can hydrogen bond to other amino acids.

*5 When two amino acids condense together, water is eliminated and a peptide link is formed. Which of the following represents this process?

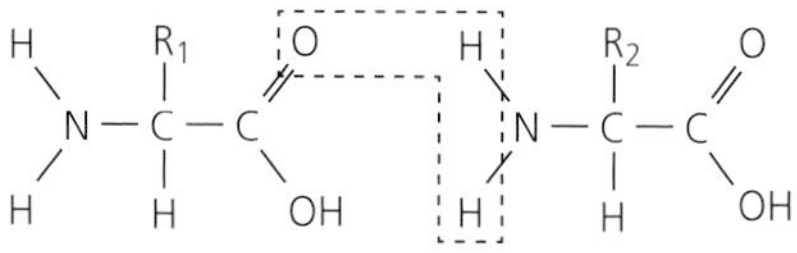

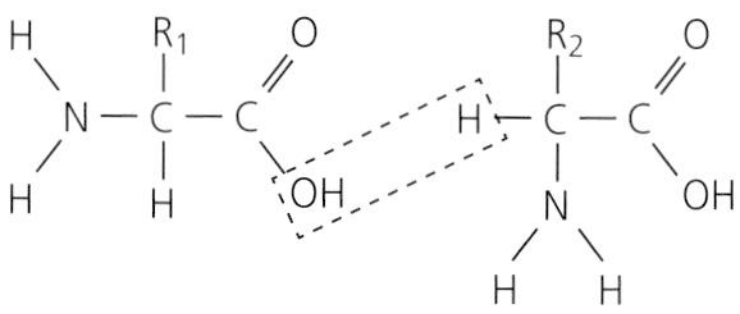

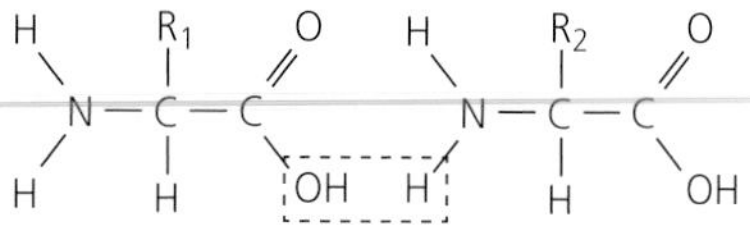

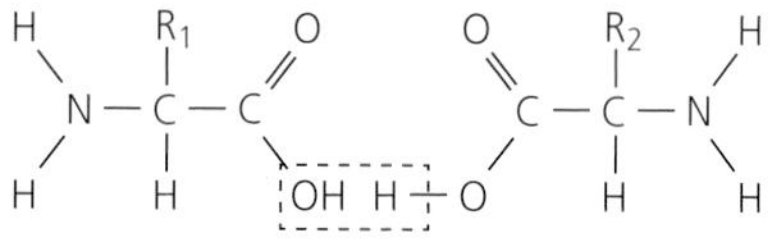

Figure 10.11

6 Which of the following statements could be used to describe what happens when a protein is denatured?

   A Covalent bonds are broken and the shape of the protein changes.

   B Intermolecular forces are broken and the peptide bonds are hydrolysed.

   C Intermolecular forces are broken and the shape of the protein changes.

   D Covalent bonds are broken and the peptide bonds are hydrolysed.

*7 Phenylalanine and alanine can react to form the dipeptide shown.

   a) On a copy of the dipeptide, circle the peptide link in the molecule.

   b) Draw a structural formula for the other dipeptide that can be formed from phenylalanine and alanine.

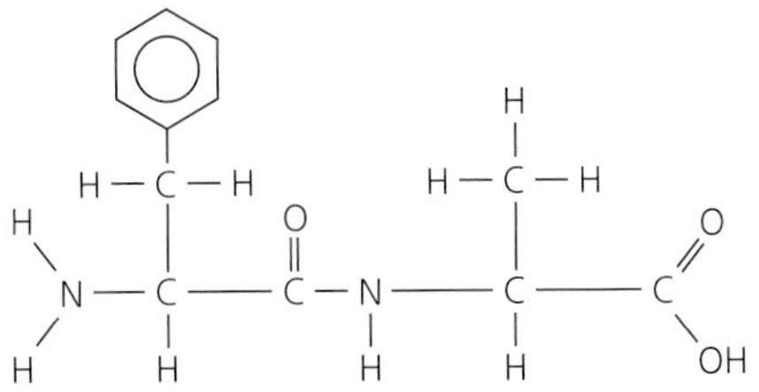

Figure 10.12

8 Collagen is a protein found in animals which helps attach muscles to the bone. Heating meat which contains collagen results in the flavour of the meat changing.

   a) With reference to collagen, explain why heating meat results in the flavour changing.

   b) Boiling meat for several hours can also result in some of the collagen hydrolysing. Name the type of compounds formed when collagen is hydrolysed.

Chapter 11
Oxidation of food

What you should know

★ Hot copper(II) oxide or acidified dichromate(VI) solutions can be used to oxidise
  ⋆ primary alcohols to aldehydes and then to carboxylic acids
  ⋆ secondary alcohols to ketones.
★ During these reactions, black copper(II) oxide forms a brown solid, and orange dichromate solution turns green.
★ Tertiary alcohols cannot be oxidised using these oxidising agents.
★ Aldehydes and ketones are molecules containing a carbonyl functional group $C=O$.
★ Straight-chain and branched aldehydes and ketones can be systematically named, and molecular formulae written, from structural formulae containing no more than eight carbons in the longest chain.
★ Aldehydes, but not ketones, can be oxidised to carboxylic acids. Oxidising agents can be used to differentiate between an aldehyde and a ketone. With an aldehyde:
  ⋆ Blue Fehling's solution forms a brick red precipitate.
  ⋆ Clear, colourless Tollens' reagent forms a silver mirror.
  ⋆ Orange acidified dichromate solution turns green.
★ For carbon compounds:
  ⋆ Oxidation is an increase in the oxygen to hydrogen ratio.
  ⋆ Reduction is a decrease in the oxygen to hydrogen ratio.
★ Many flavour and aroma molecules are aldehydes.
★ Oxygen from the air causes the oxidation of food. The oxidation of edible oils gives food a rancid flavour.
★ Antioxidants
  ⋆ are molecules that prevent unwanted oxidation reactions occurring
  ⋆ are substances that are easily oxidised, and oxidise in place of the compounds they have been added to protect
  ⋆ can be identified as the substance being oxidised in a redox equation.

Oxidation of alcohols

Primary and secondary alcohols can be oxidised by various oxidising agents but tertiary alcohols do not undergo **oxidation** readily. Suitable oxidising agents are *acidified potassium dichromate* and *hot copper (II) oxide*.

Remember

- When primary alcohols are oxidised they produce aldehydes.
- When secondary alcohols are oxidised they produce **ketones**.

Table 11.1

Oxidising agent	Observations
Copper (II) oxide	A brown solid is formed from the black copper(II) oxide
Acidified potassium dichromate	Change in colour noted from orange to green

The following tables show the structures of aldehydes and ketones and their parent alcohols.

Table 11.2 Oxidation of ethanol produces ethanal.

Alcohol	Aldehyde
Ethanol	Ethanal
CH_3CH_2OH	CH_3CHO
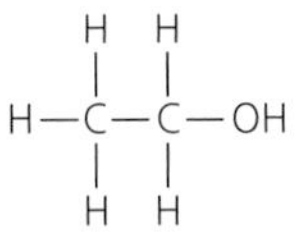 The structure of ethanol	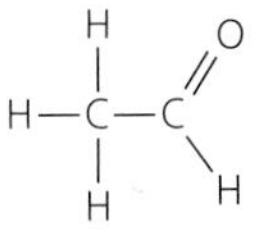 The structure of ethanal

Table 11.3 Oxidation of propan-1-ol produces propanal.

Alcohol	Aldehyde
Propan-1-ol	Propanal
$CH_3CH_2CH_2OH$	CH_3CH_2CHO
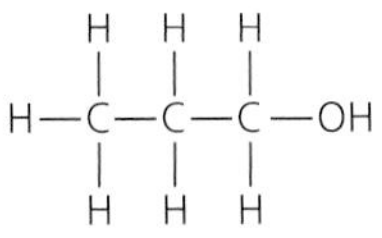 The structure of propan-1-ol	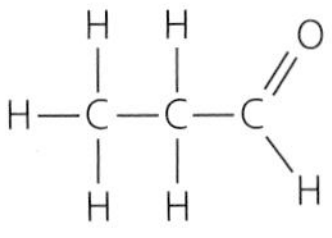 The structure of propanal

Table 11.4 Oxidation of propan-2-ol produces propanone.

Alcohol	Ketone
Propan-2-ol	Propanone
$CH_3CH(OH)CH_3$	CH_3COCH_3
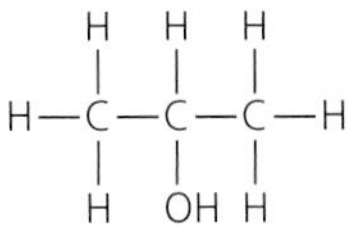 The structure of propan-2-ol	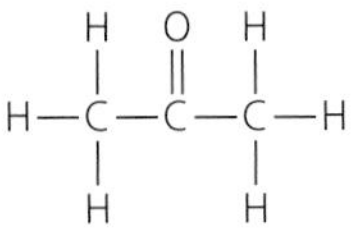 The structure of propanone

Table 11.5 Oxidation of butan-2-ol produces butanone.

Alcohol	Ketone
Butan-2-ol	Butanone
$CH_3CH_2CH(OH)CH_3$	$CH_3CH_2COCH_3$
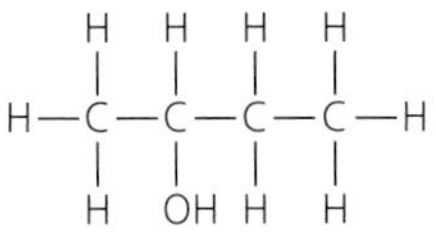 The structure of butan-2-ol	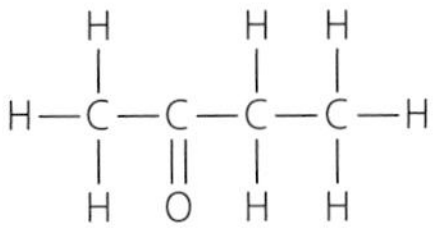 The structure of butanone

Aldehydes and ketones can be identified by the presence of the carbon–oxygen double bond, C=O, which is known as the **carbonyl group**. In an aldehyde, the carbonyl group is at the end of the carbon chain and has a hydrogen atom attached to it. In a ketone, the carbonyl group is joined to two other carbon atoms and does not have a hydrogen atom attached to it.

Branched aldehydes and ketones are named by giving priority to the carbonyl functional group.

Examples are shown in Table 11.6.

Table 11.6 Naming aldehydes and ketones

Structure	Name	Aldehyde or ketone
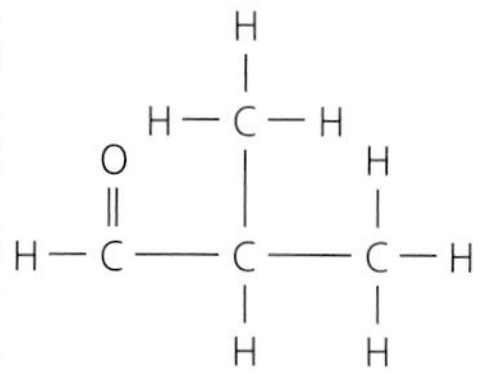	2-Methylpropanal	Aldehyde
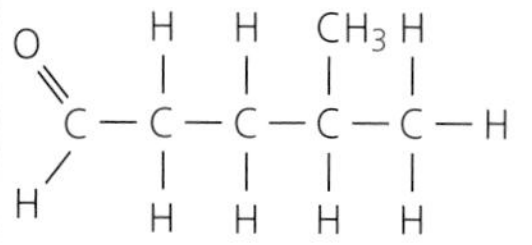	4-Methylpentanal	Aldehyde
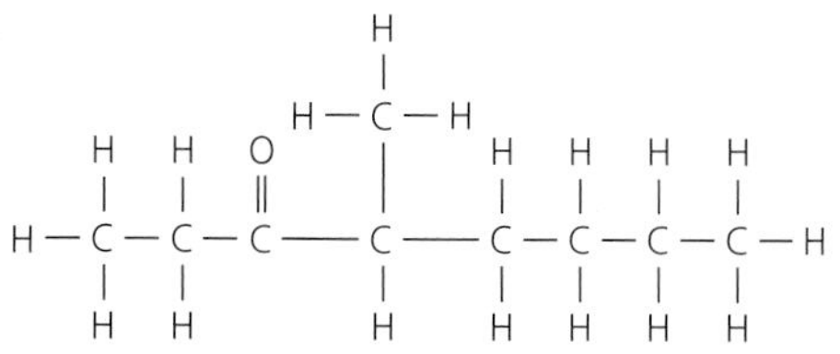	4-Methyloctan-3-one	Ketone
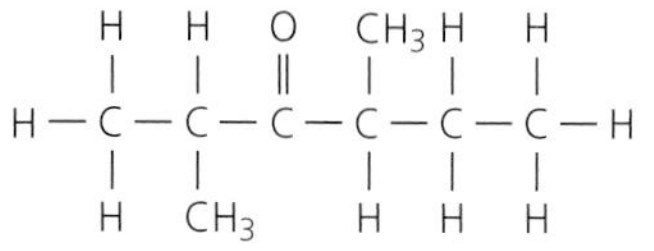	2,4-Dimethylhexan-3-one	Ketone

Hints & tips

There is no need to indicate a number for the functional group when naming an aldehyde as the carbonyl group is always at the end of the carbon chain. However, when naming a ketone, it is usually necessary to specify the number of the carbonyl group.

Oxidation of aldehydes

Oxidising agents can be used to distinguish between aldehydes and ketones since aldehydes can oxidise but ketones cannot. The three most common oxidising agents used are *Fehling's solution, Tollen's reagent* and *acidified potassium dichromate*.

Table 11.7 Oxidising agents

Oxidising agent	Observations	Explanation
Acidified potassium dichromate solution	Orange → green	$Cr_2O_7^{2-}(aq)$ reduced to $Cr^{3+}(aq)$
Fehling's solution	Blue → orange/red	$Cu^{2+}(aq)$ reduced to $Cu_2O(s)$ i.e. $Cu^{2+} + e^- \rightarrow Cu^+$
Tollen's reagent	Colourless → silver	$Ag^+(aq)$ reduced to $Ag(s)$ i.e. $Ag^+ + e^- \rightarrow Ag$

When oxidised, aldehydes form carboxylic acids, as shown in Tables 11.8 and 11.9.

Table 11.8 Oxidation of ethanal produces ethanoic acid.

Aldehyde	Carboxylic acid
Ethanal	Ethanoic acid
CH_3CHO	CH_3COOH
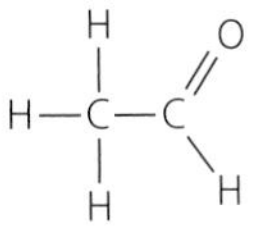 The structure of ethanal	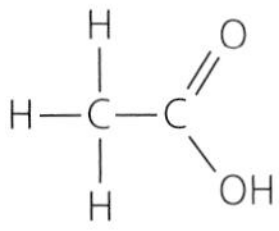 The structure of ethanoic acid

Table 11.9 Oxidation of butanal produces butanoic acid.

Aldehyde	Carboxylic acid
Butanal	Butanoic acid
$CH_3CH_2CH_2CHO$	$CH_3 CH_2CH_2COOH$
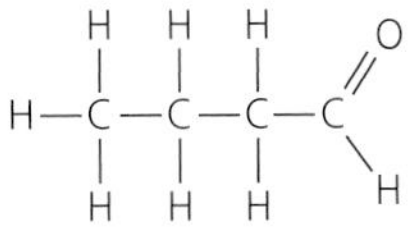 The structure of butanal	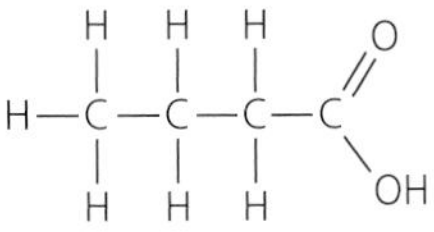 The structure of butanoic acid

Carboxylic acids contain the carboxyl functional group (–COOH).

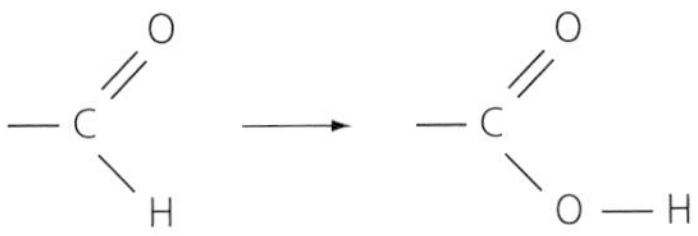

Figure 11.1 Oxidation of an aldehyde produces a carboxylic acid.

Hints & tips

The oxidation of alcohols can be summarised as follows:

1 *Primary alcohol →
Aldehyde →
Carboxylic acid*

2 *Secondary alcohol →
Ketone (not readily
oxidised)*

3 *Tertiary alcohol (not
readily oxidised)*

Oxidation and the oxygen to hydrogen ratio

When dealing with carbon compounds, a useful method for determining whether oxidation or reduction has occurred involves calculating the oxygen to hydrogen ratio. This is illustrated in Table 11.10, which shows the oxidation of ethanol to form ethanal and then ethanoic acid.

Table 11.10 Oxidation and the O:H ratio

Primary alcohol	Aldehyde	Carboxylic acid
Ethanol	Ethanal	Ethanoic acid
CH_3CH_2OH	CH_3CHO	CH_3COOH
O:H ratio 1:6	O:H ratio 1:4	O:H ratio 1:2

We can use the oxygen to hydrogen ratio to show that:
- oxidation occurs when there is an increase in the oxygen to hydrogen ratio
- reduction occurs when there is a decrease in the oxygen to hydrogen ratio.

Reactions of carboxylic acids
Reduction

Carboxylic acids are produced from the oxidation of primary alcohols and aldehydes. The reverse of this reaction can be used to produce primary alcohols and aldehydes. In other words, carboxylic acids can undergo **reduction**. This is illustrated in Figure 11.2.

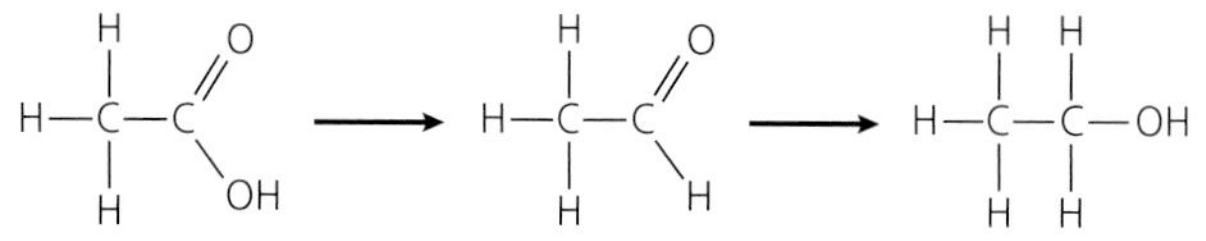

Figure 11.2 Reduction of ethanoic acid

Flavour

Different cooking methods can alter the flavour and nutritional value of a food by affecting the chemical compounds responsible for flavour and nutrition. Vitamin C, shown in Figure 11.3, contains a number of polar –OH groups which can hydrogen bond to water. As a result, vitamin C is soluble in water so cooking foods containing this vitamin in water can rob the food of its valuable vitamin C content since the vitamin C will dissolve in the cooking water, which is usually discarded.

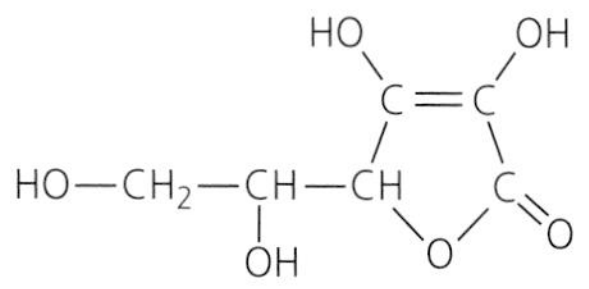

Figure 11.3 Vitamin C is a water-soluble vitamin.

Figure 11.4 Limonene

Other molecules in food are non-polar and do not, therefore, dissolve in water. Limonene is a non-polar molecule responsible for the lemon flavour in lemons. When lemons are cooked in oil, which is also non-polar, the limonene molecules will dissolve in the oil. This can be useful for creating oils with a pleasant lemon taste.

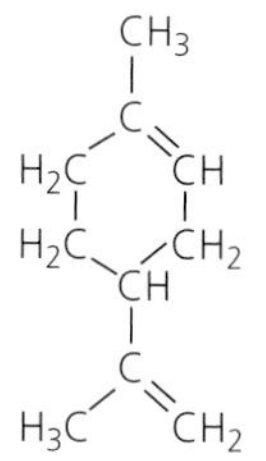

Heating foods usually alters the taste as the chemical structure of flavour molecules is changed in some way. For example, frying an egg leads to an obvious chemical reaction as a solid 'egg white' is formed. Egg whites are a great source of protein. The protein molecules, which contain polar –NH and –C$=$O groups, bind to other protein molecules through hydrogen bonding which gives them a specific shape. Heating an egg breaks these hydrogen bonds causing a complete change of shape. This process of the protein changing shape is known as **denaturing**. Whenever a protein-containing food is heated, denaturing can occur resulting in a change of flavour. Another example is the cooking of meat, which has a particularly high protein content. The denaturing of the proteins in meat as it cooks results in some of the undesirable toughness being lost, making the meat more palatable.

Hints & tips

When looking at unfamiliar molecules, look for the obvious functional groups that will help you decide whether the molecule is polar or non-polar:

- *Hydrocarbon structures are likely to suggest that the molecule is non-polar.*
- *Hydroxyl (–OH), carbonyl (–C=O), carboxyl (–COOH), amino (–NH_2) and amide (–CONH) groups are all polar.*

Remember, polar molecules will dissolve in water; non-polar molecules will dissolve in oil. In addition, polar molecules are likely to have higher boiling points than non-polar molecules.

Antioxidants

Many recognisable flavour and aroma molecules are aldehydes and ketones, as shown in Table 11.11 and Figure 11.5.

Table 11.11 Aldehydes and ketones are flavour molecules.

Compound name	Where found
Cinnamaldehyde	Cinnamon
Furfural	Coffee
2,3-Butanedione	Butter
1-(2-Pyridinyl)ethanone	Popcorn
Heptan-2-one	Blue cheese

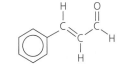

Figure 11.5 Cinnamaldehyde

As we now know, aldehydes can undergo oxidation to produce carboxylic acids. When this occurs in a food, the flavour and aroma can change. Quite often, this results in the food becoming spoiled as the carboxylic acids produced give a very acidic taste. Consequently, food chemists have developed methods for preventing the oxidation of foods in an attempt to preserve the flavour. A common method of preventing oxidation involves the addition of an **antioxidant** to the food. One of the most common antioxidants used is ascorbic acid, the chemical name for vitamin C. Ascorbic acid acts as an antioxidant as it will readily undergo oxidation itself and thus save the food from becoming oxidised. The **ion–electron equation** for the oxidation of ascorbic acid is:

$$C_6H_8O_6 \rightarrow C_6H_6O_6 + 2H^+ + 2e^-$$

Another common problem which results in food spoilage is the oxidation of edible oils. Reacting oils with oxygen significantly alters the taste of oil-containing foods. Food manufacturers will often package such foods in an atmosphere of nitrogen to minimise exposure to oxygen. Alternatively, the addition of antioxidant compounds helps to prevent the edible oils from going rancid through oxidation.

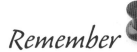

Remember

Antioxidants are molecules that:

- prevent unwanted oxidation reactions occurring
- will oxidise in place of the compounds they have been added to protect.

Key points

* Primary and secondary alcohols can be oxidised to form aldehydes and ketones using oxidising agents such as acidified potassium dichromate or hot copper (II) oxide.
* Tertiary alcohols cannot be oxidised using mild oxidising agents.
* Aldehydes and ketones contain the carbonyl group (–C=O).
* Aldehydes can be oxidised further to produce carboxylic acids using acidified potassium dichromate, Fehling's solution or Tollen's reagent; ketones cannot be oxidised.
* Carboxylic acids can be reduced to form aldehydes or primary alcohols.
* Oxidation involves an increase in the O:H ratio.
* Flavour and aroma molecules in food are often aldehydes and ketones. Oxidation of these molecules can change the flavour, making foods taste unpleasant, so antioxidants are used to prevent this oxidation and thus preserve flavour.
* The chemical structure of a food molecule can be used to determine whether the molecule is likely to be soluble or insoluble in water or oil. The volatility of a flavour compound is also determined by its structure.

Study questions

Questions 1–5 refer to the molecules shown in Table 11.12.

Table 11.12

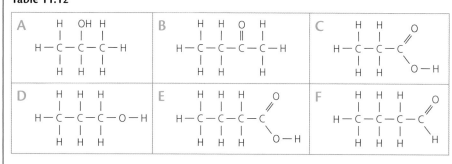

1 Identify the compound which can be oxidised to form the compound shown in box C.

2 Identify the compound which can be formed when the compound shown in box E is reduced.

3 Identify the secondary alcohol.

4 Identify the ketone.

5 Which two compounds could react with NaOH(aq) to form a salt?

6 Compound Y was oxidised to produce a compound which did not react with sodium hydroxide. Compound Y could be

   A propanone   C propanal
   B propanoic acid   D propan-1-ol.

7 Vanillin and eugenol are examples of small molecules which are responsible for the flavour in food. Their structures are shown in Figures 11.6 and 11.7. With reference to the structures, explain why vanillin is more soluble in water than eugenol.

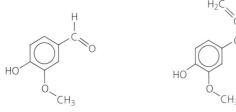

Figure 11.6 Vanillin   **Figure 11.7** Eugenol

8 State the systematic names of the following compounds:

a) 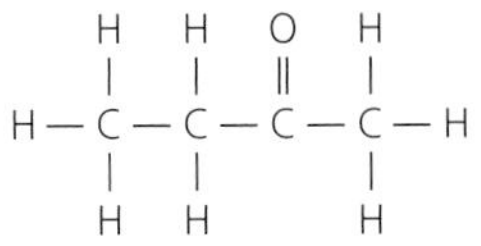
$$H-\underset{\underset{H}{|}}{\overset{\overset{H}{|}}{C}}-\underset{\underset{H}{|}}{\overset{\overset{H}{|}}{C}}-\overset{\overset{O}{\|}}{C}-\underset{\underset{H}{|}}{\overset{\overset{H}{|}}{C}}-H$$

c)
$$H-\underset{\underset{H}{|}}{\overset{\overset{H}{|}}{C}}-\underset{\underset{H}{|}}{\overset{\overset{H}{|}}{C}}-\underset{\underset{H}{|}}{\overset{\overset{H}{|}}{C}}-\underset{\underset{H}{|}}{\overset{\overset{H}{|}}{C}}-\overset{O}{\underset{H}{C}}$$

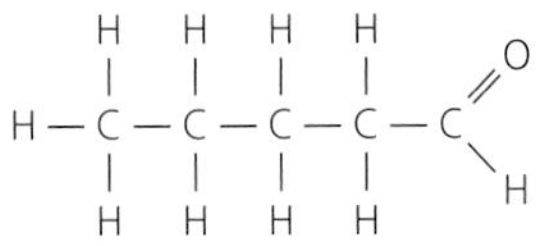

b) 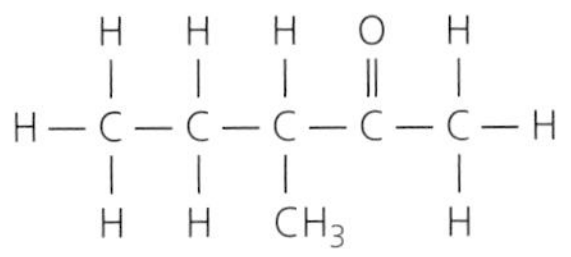
$$H-\underset{\underset{H}{|}}{\overset{\overset{H}{|}}{C}}-\underset{\underset{H}{|}}{\overset{\overset{H}{|}}{C}}-\underset{\underset{CH_3}{|}}{\overset{\overset{H}{|}}{C}}-\overset{\overset{O}{\|}}{C}-\underset{\underset{H}{|}}{\overset{\overset{H}{|}}{C}}-H$$

d) 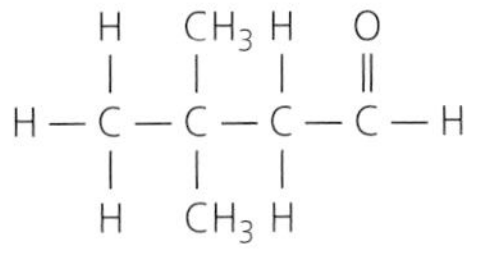
$$H-\underset{\underset{H}{|}}{\overset{\overset{H}{|}}{C}}-\underset{\underset{CH_3}{|}}{\overset{\overset{CH_3}{|}}{C}}-\underset{\underset{H}{|}}{\overset{\overset{H}{|}}{C}}-\overset{\overset{O}{\|}}{C}-H$$

Figure 11.8

9 a) For the compounds shown in Question 8(a) and 8(b), draw full structural formulae for the compounds formed when these compounds are reduced.
   b) For the compounds shown in Question 8(c) and 8(d), draw full structural formulae for the compounds formed when these compounds are oxidised.

10 a) Draw the structures of butan-1-ol and butan-1,3-diol.
   b) Explain why butan-1,3-diol has a higher boiling point than butan-1-ol.

Chapter 12
Fragrances

What you should know

★ Essential oils are concentrated extracts of the volatile, non-water-soluble aroma compounds from plants. They are mixtures of many different compounds. They are widely used in perfumes, cosmetic products, cleaning products and as flavourings in foods.
★ Terpenes are key components in most essential oils. They are unsaturated compounds formed by joining together isoprene (2-methylbuta-1,3-diene) units.
★ Terpenes can be oxidised within plants to produce some of the compounds responsible for the distinctive aromas of spices.
★ Given the structural formula for a terpene-based molecule
    ★ an isoprene unit can be identified within the molecule
    ★ the number of isoprene units joined together within the molecule can be stated.

Terpenes

Essential oils are concentrated extracts of the aroma compounds found in plants. Essential oils are widely used in perfumes, cosmetic products, cleaning products and as flavourings. Some examples are shown in Table 12.1.

Table 12.1 Examples of essential oils

Name of essential oil	Use
Citronella oil	Insect repellant
Frankincense oil	Perfumes
Spearmint oil	Flavouring in mouthwashes
Orange oil	Cleaning products

Essential oils are not the same as edible oils. Essential oils contain many compounds that are usually volatile and insoluble in water. An example of a family of compounds found in essential oils is the **terpene**.

Terpenes are compounds based on isoprene (2-methylbuta-1,3-diene), which has the molecular formula C_5H_8.

All terpenes contain isoprene units joined together. Terpene compounds can be given the formula $(C_5H_8)n$, where n is the number of isoprene units joined together. For example, the terpene known as myrcene is formed from two isoprene units joined together, as shown in Figure 12.2.

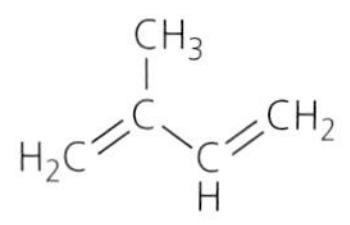

Figure 12.1 The structure of isoprene

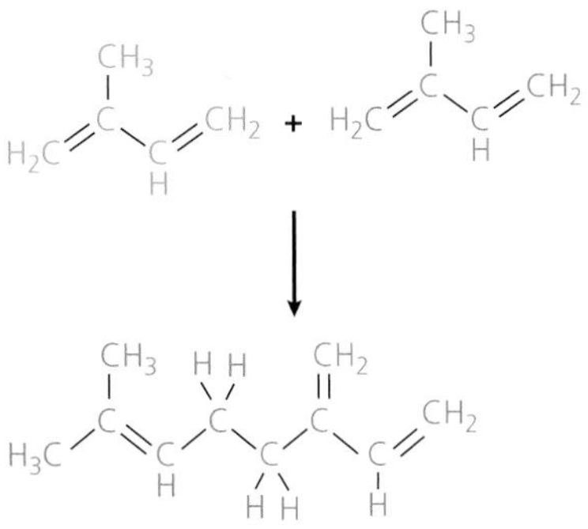

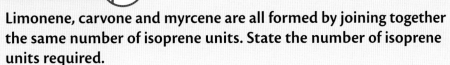

Figure 12.2 Myrcene is formed by joining together two isoprene units. Note the two units highlighted

Limonene, carvone and myrcene are all examples of terpenes formed from two isoprene units joining together.

Table 12.2 Examples of terpenes found in essential oils

Limonene	Carvone	Myrcene
CH$_3$ structure	CH$_3$ structure	CH$_2$ structure
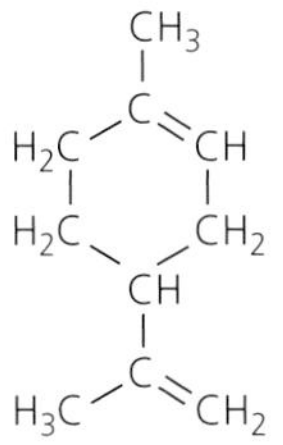 Citrus	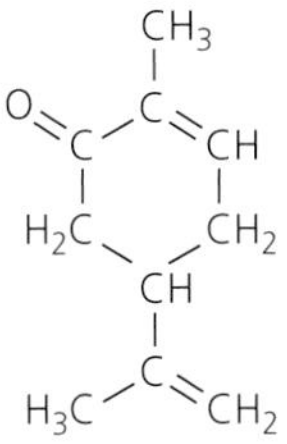 Spearmint	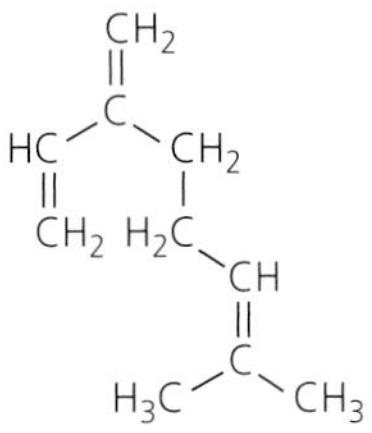 Woody smell

Example

Limonene, carvone and myrcene are all formed by joining together the same number of isoprene units. State the number of isoprene units required.

Solution

To solve this, add up the number of carbon atoms in the structure and divide by 5. For example, limonene has 10 carbon atoms. This means that it has been formed from two isoprene units.

Carvone and myrcene also have 10 carbon atoms. Therefore, the answer is 2.

Oxidation of terpenes

In nature, the oxidation of terpenes produces many of the compounds responsible for the aroma of spices derived from plants. For example, peppermint oil contains the terpene menthol and its oxidation product, menthone. Both compounds contribute to the flavour and aroma of the oil. An examination of the structures of menthol and menthone shows that the structural change is an example of a secondary alcohol (menthol) being oxidised to form a ketone (menthone).

Table 12.3 Oxidation of menthol produces menthone.

Terpene	Oxidised terpene	Where found?
Menthol	Menthone	Peppermint oil

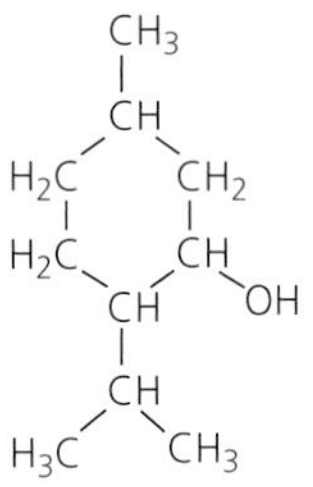

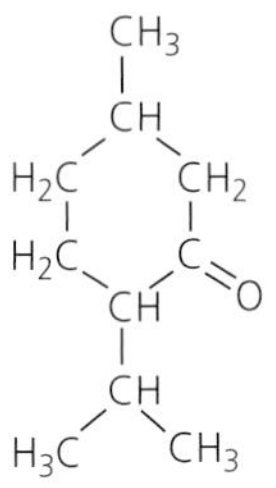

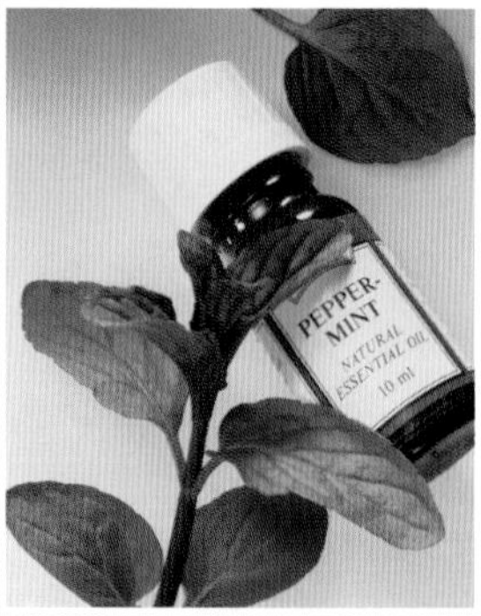

Study questions ?

1 Farnesol is a terpene found in the essential oil derived from a rose.

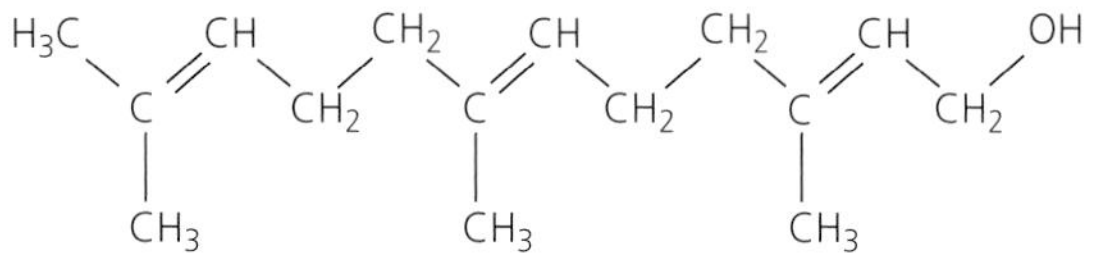

Figure 12.3

   a) Terpenes are based on isoprene. Draw a structural formula for isoprene.
   b) How many isoprene units must join to form farnesol?

2 a) Squalene is a terpene found in shark oil. Describe how a solution of bromine water could be used to distinguish between squalene and farnesol.
   b) State the number of isoprene units which join to form squalene.

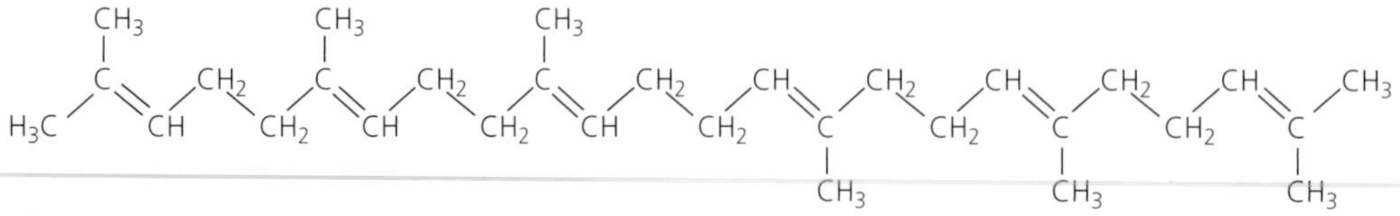

Figure 12.4

3 Citral (Figure 12.5) is a terpene oxidation product found in lemongrass. It is formed from the oxidation of nerol. Menthone (Figure 12.6) is a terpene oxidation product found in peppermint oil.

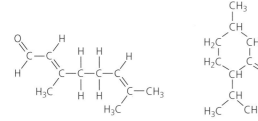

Figure 12.5 Citral   **Figure 12.6** Menthone

a) Describe an experiment you could carry out that would allow you to distinguish between citral and menthone.

b) Draw a structural formula for the terpene nerol.

*4 Erythrose (Figure 12.7) can be used in the production of a chewing gum that helps prevent tooth decay. Which of the compounds shown in Figure 12.8 will be the *best* solvent for erythrose?

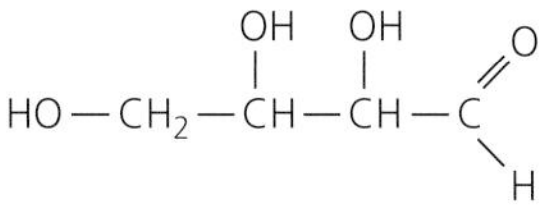

$$HO-CH_2-CH-CH-C \overset{O}{\underset{H}{\big\backslash}}$$
$$\overset{|}{OH} \quad \overset{|}{OH}$$

Figure 12.7 Erythrose

A
$$CH_2$$
$$H_2C \qquad CH_2$$
$$| \qquad |$$
$$H_2C \qquad CH_2$$
$$CH_2$$

B $CH_3-CH_2-CH_2-CH_2-CH_2-CH_3$

C CH_3-CH_2-OH

D
$$Cl-\overset{\overset{H}{|}}{\underset{\underset{Cl}{|}}{C}}-Cl$$

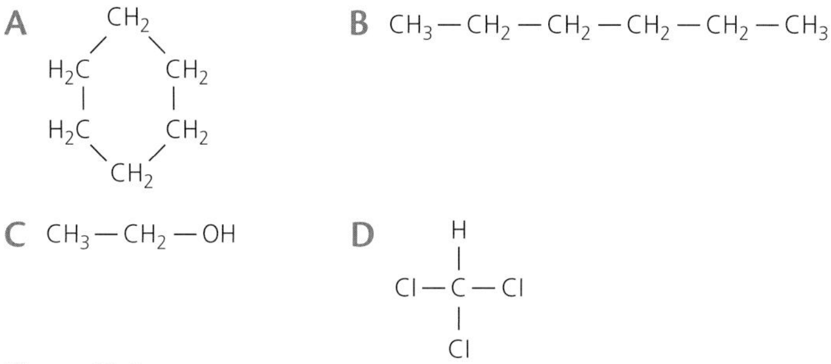

Figure 12.8

5 Limonene has the structure shown in Figure 12.9.

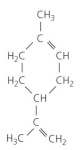

Figure 12.9 Limonene

Which of the following statements cannot be applied to limonene?

A It is a saturated terpene.
B It contains two isoprene units.
C It can undergo addition reactions.
D It satisfies the formula $(C_5H_8)_n$.

*6 Myrcene (Figure 12.10) is a simple terpene. Terpenes contain at least one isoprene unit. Which of the structures given in Figure 12.11 shows a correctly highlighted isoprene unit?

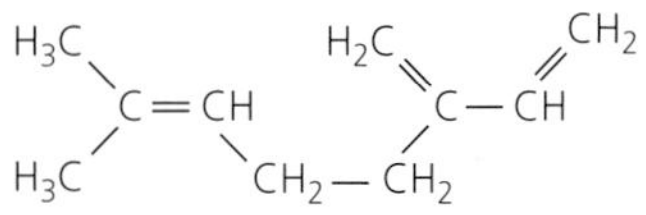

Figure 12.10 Myrcene

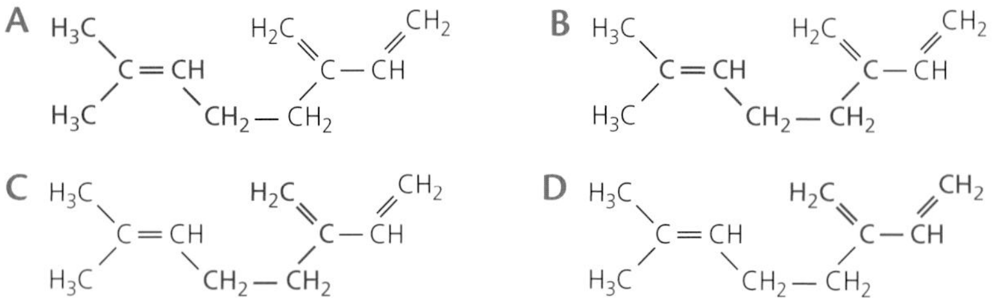

Figure 12.11

*7 Two typical compounds that are present in many perfumes are shown.

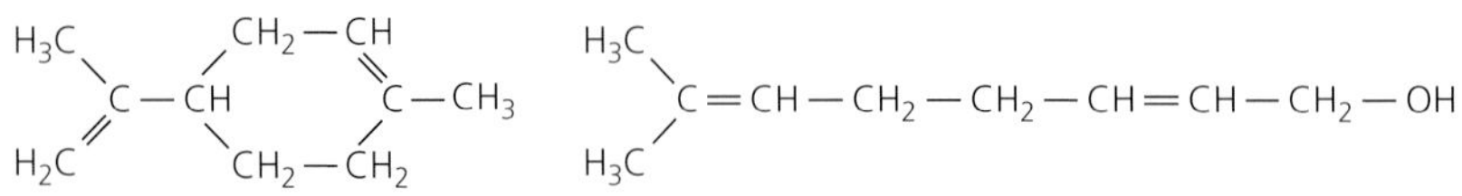

$C_{10}H_{16}$
Limonene

$C_9H_{16}O$
Geraniol

Figure 12.12

a) Why does geraniol evaporate more slowly than limonene?

b) The structure of one of the first synthetic scents used in perfume is shown below.

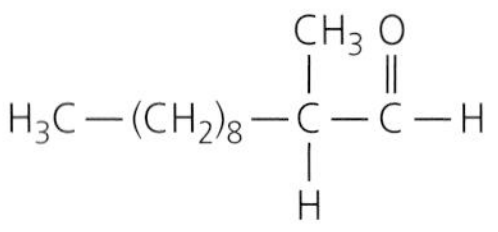

Figure 12.13

i) Name the family of carbonyl compounds to which this synthetic scent belongs.

ii) Copy and complete the structure below to show the product formed when this scent is oxidised.

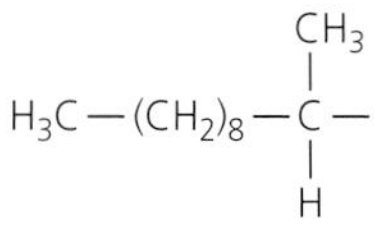

Figure 12.14

Ultraviolet light (UV)

Sunlight contains **ultraviolet light**, which is a very high energy light capable of breaking chemical bonds. Exposing our skin to UV light is beneficial as it allows us to make vitamin D, but it also causes our skin to age, and too much UV exposure can cause sunburn. As sunburn has been linked to skin cancer, chemists have developed products to protect our skin from the damaging effects of UV. *Sunscreens* contain compounds that filter the UV light so that less UV reaches the skin. *Sunblock* contains compounds that reflect the UV so that it does not reach the skin at all.

Free radicals

UV light has enough energy to break covalent bonds leaving two atoms with unpaired electrons. For example, a chlorine molecule (Cl_2) contains two chlorine atoms bonded together by a covalent bond as shown in Figure 13.1.

When UV light is shone onto chlorine, the energy supplied causes the covalent bond to break to form two chlorine atoms. These atoms are highly reactive as they each have an unpaired electron. They are known as **free radicals**.

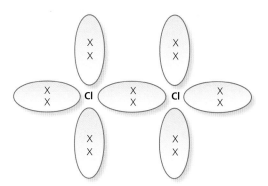

Figure 13.1 Two chlorine atoms joined by a covalent bond

This chlorine atom is highly reactive as it has an unpaired electron. It will react rapidly with any other atom or molecule that comes into contact with it.

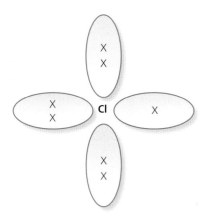

Figure 13.2 A chlorine free radical. Note the unpaired electron

Free radical reactions

Chemists have studied reactions involving free radicals and have observed a particular sequence of reactions. This is best illustrated by describing the reaction of methane, an alkane, with chlorine.

The overall reaction is:

$$CH_4 + Cl_2 \rightarrow CH_3Cl + HCl$$

This is known as a substitution reaction as one of the H atoms in the methane is substituted for a Cl atom. There are three steps to this reaction: **initiation**, **propagation** and **termination**.

Initiation

UV light is shone on a mixture of methane and chlorine. There is enough energy in the UV light to break the Cl–Cl bond to produce two Cl atoms.

$$Cl - Cl \rightarrow Cl^\bullet + Cl^\bullet$$

Propagation

The highly reactive chlorine radicals ($Cl^\bullet$) react with the methane molecules. This produces methyl radicals which then react with chlorine molecules (Cl_2).

$$Cl^\bullet + CH_4 \rightarrow CH_3^\bullet + HCl$$

$$CH_3^\bullet + Cl_2 \rightarrow CH_3Cl + Cl^\bullet$$

Termination

Two free radicals combine to form stable molecules.

$$Cl^\bullet + Cl^\bullet \rightarrow Cl_2$$

$$CH_3^\bullet + CH_3^\bullet \rightarrow C_2H_6$$

$$CH_3^\bullet + Cl^\bullet \rightarrow CH_3Cl$$

Example

Write initiation, propagation and termination steps for the reaction between bromine and propane.

Solution

Initiation:

$Br - Br \rightarrow Br^{\bullet} + Br^{\bullet}$

Propagation:

$Br^{\bullet} + C_3H_8 \rightarrow C_3H_7^{\bullet} + HBr$

$C_3H_7^{\bullet} + Br_2 \rightarrow C_3H_7Br + Br^{\bullet}$

Termination:

$Br^{\bullet} + Br^{\bullet} \rightarrow Br_2$

$C_3H_7^{\bullet} + C_3H_7^{\bullet} \rightarrow C_6H_{14}$

$C_3H_7^{\bullet} + Br^{\bullet} \rightarrow C_3H_7Br$

Free radical scavengers

Given that UV light can cause free radicals to form on our skin, which leads to ageing, cosmetic chemists have developed compounds that combine with these radicals to form stable molecules. These compounds are known as **free radical scavengers**. As they form stable molecules, they help to stop the free radical chain reactions that cause the skin to form wrinkles. Examples of free radical scavengers include vitamins C and E. Free radical scavengers are also added to food and plastics to prevent spoiling or damage by free radicals.

Key points

* UV light is a high energy form of light that can break chemical bonds.
* Exposure to UV light causes ageing of the skin and can cause sunburn.
* Sunblocks stop UV light reaching the skin.
* Free radicals are highly reactive atoms or molecules with unpaired electrons.
* Free radicals can take part in chemical reactions that involve the steps initiation, propagation and termination.
* Free radical scavengers are molecules that react with free radicals to prevent chain reactions.
* Free radical scavengers are added to cosmetic products such as skin creams. They are also added to foods and plastics.

Study questions

1 **Table 13.1**

A	B	C
$F_2 \rightarrow F^{\bullet} + F^{\bullet}$	$Cl^{\bullet} + Cl^{\bullet} \rightarrow Cl_2$	$CH_3^{\bullet} + Cl_2 \rightarrow CH_3Cl + Cl^{\bullet}$
D	E	F
$CH_3^{\bullet} + CH_3^{\bullet} \rightarrow C_2H_6$	$F_2 + H_2 \rightarrow 2HF$	$C_2H_4 + Br_2 \rightarrow C_2H_4Br_2$

a) Which box shows an initiation reaction?
b) Which two boxes show termination reactions?
c) Which box shows a propagation reaction?

*2 Suncreams contain antioxidants. The antioxidant, compound A, can prevent damage to skin by reacting with free radicals such as $NO_2^{\bullet}$. Why can compound A be described as a free radical scavenger in the reaction shown below?

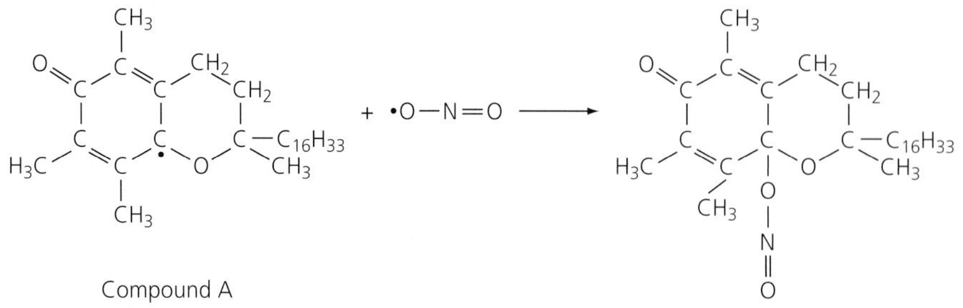

Compound A

Figure 13.3

3 Fluorine can react with ethane in a free radical reaction. The overall reaction is:

$$F_2 + C_2H_6 \rightarrow C_2H_5F + HF$$

a) Explain why UV light can start this reaction.

b) Why should we limit our exposure to UV light?

c) What is meant by the term 'free radical'?

d) Write equations for the initiation, propagation and termination steps.

e) Fluorine can react with ethene without the need for UV light to start the reaction. Explain this observation.

Section 3 Chemistry in Society

Chapter 14

Getting the most from reactants

What you should know

★ Industrial processes are designed to maximise profit and minimise the impact on the environment.
★ Factors influencing industrial process design include:
  ★ availability, sustainability and cost of feedstock(s)
  ★ opportunities for recycling
  ★ energy requirements
  ★ marketability of by-products
  ★ product yield.
★ Environmental considerations include:
  ★ minimising waste
  ★ avoiding the use or production of toxic substances
  ★ designing products which will biodegrade if appropriate.

Designing an industrial process

The chemical industry manufactures a huge variety of products including medicines, plastics, paints and cosmetics. As with any business, the chemical industry makes these products to generate a profit but it must consider how to make maximum profit while having minimal impact on people and the environment. Table 14.1 shows some of the factors that could influence the design of an industrial process and their effects on profits.

Table 14.1 Factors affecting the design of a chemical process

Factor	Profit making	Profit losing
Availability, sustainability and cost of **feedstock**(s)*	If the process relies on a feedstock which is available locally, this will help to keep costs down and maximise profit.	If the feedstock has to be transported from further afield, this will have significant cost, safety and environmental considerations. Is the feedstock likely to last for a long time or is it likely to become scarce? This could significantly affect the cost of buying the feedstock since rare materials cost more. If the feedstock is too expensive, alternatives might have to be investigated. Energy derived from oil and gas fluctuates in price, but always becomes expensive in times of tension in the producing areas.

Table 14.1 (continued)

Factor	Profit making	Profit losing
Opportunities for recycling	If unreacted starting materials can be fed back into the chemical reactor to form new products, this will improve the efficiency and profitability of the process. If water used in the process can be recycled, this reduces waste.	If it is very difficult to separate unreacted starting materials at the end of reaction, this makes the reaction inefficient and wasteful.
Energy requirements	Exothermic reactions can be used to sustain the heat in a reaction or heat the building. This saves money on energy costs. Many chemical reactions use catalysts to speed up the rate of reaction rather than using higher temperatures. This can allow more control of the reaction products and saves energy.	Reactions which require heating can be very costly as energy (gas, electricity) must be purchased. Reactions which require cooling can also be expensive as energy must be removed from the reaction by surrounding with a coolant.
Use of by-products	Many reactions produce more than one product. If the other product(s) can be used elsewhere in the process, this will save money. If the by-products can be sold to other companies, this will also increase profit.	If the by-products are toxic, very corrosive or environmentally damaging, it will be expensive to deal with these. For example, acidic gases like sulfur dioxide (SO_2) and the greenhouse gas carbon dioxide (CO_2) can be costly for companies to deal with.
Yield of product	A high yield of product is very profitable.	Low yields cost money as time and energy must be put in to run the reaction several times to produce enough of the desired product.

*A feedstock is a substance obtained from a **raw material** that is used to make another substance. For example, air is a raw material from which oxygen gas can be obtained. Oxygen is the feedstock.

Environmental considerations

Every industry has a duty to protect the people who make its products, use its products and live nearby. The chemical industry has worked hard to promote a culture of health and safety and minimal impact on the environment.

Once a product has been identified, chemists and engineers will work together to address these considerations. Sometimes this will involve attempting to make a new compound using an alternative route that does not generate so much waste or investigating new types of catalyst to make a reaction more efficient. Recycling of reactants is a common and obvious step taken to reduce waste in a chemical process.

Remember

The three main environmental considerations of the chemical industry are:
1 How can waste be minimised?
2 How can we avoid making toxic substances?
3 Can we design products that will biodegrade?

Key points !

* Industrial processes are designed to maximise profit and minimise the impact on the environment.
* Factors that influence the design of a chemical process include: availability, sustainability and cost of feedstocks; opportunities for recycling; energy requirements; use of by-products; product yield.
* Environmental considerations include: minimising waste; avoiding the use or production of toxic substances; designing products which will biodegrade.

Hints & tips ⭐

It is common for scientists and engineers to represent chemical reactions using flow diagrams. An example is shown in Figure 11.1, which shows how ethanol is made by reacting ethene with water.

You should be able to look at these diagrams and apply your chemical knowledge.

1 *Look out for opportunities to **recycle**. In this case, unreacted ethene from separator A could be fed back into the reaction vessel.*

2 *Look out for evidence of **endo- or exothermic reactions**. In this case, a cooler is used suggesting that the reaction between ethene and water is exothermic. This has a cost consideration as cooling (and heating) can be expensive.*

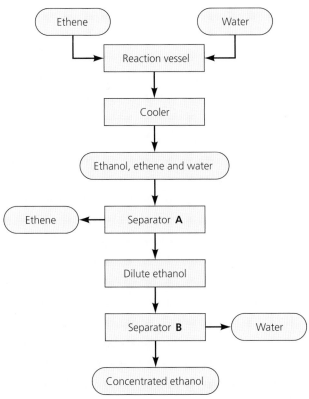

Figure 14.1 Representing a chemical reaction using a flow diagram

3 *Think about how chemicals can be **separated**. Solids can be removed from liquids using **filtration**, and **distillation** can be used for separating a mixture of liquids. In this case, it is likely that distillation is the process used in separator B to separate water and ethanol.*

4 *Look for **environmentally damaging** products or **by-products**. In this case, water is produced from separator B. It would make sense to feed this water back into the reaction vessel.*

Study questions

1 Hydrogen gas can be made on an industrial scale using the sulfur–iodine cycle. There are three steps in the sulfur–iodine cycle:

Step 1: $I_2 + SO_2 + 2H_2O \rightarrow 2HI + H_2SO_4$

Step 2: $2HI \rightarrow I_2 + H_2$

Step 3: $H_2SO_4 \rightarrow SO_2 + H_2O + \frac{1}{2}O_2$

  a) Suggest why care must be taken to ensure the products of step 3 are not released into the atmosphere.

  b) Why does step 3 help to reduce the cost of manufacturing hydrogen?

  c) Write the overall equation for the sulfur–iodine cycle.

*2 Urea, $(NH_2)_2CO$, is an important industrial chemical that is mainly used in fertilisers. It is made by the Bosch–Meiser process.

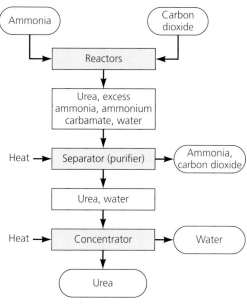

Figure 14.2

  a) In the reactors, the production of urea involves two reversible reactions.
In the first reaction ammonium carbamate is produced.

$$2NH_3(g) + CO_2(g) \rightleftharpoons H_2NCOONH_4(g)$$

In the second reaction the ammonium carbamate decomposes to form urea.

$$H_2NCOONH_4(g) \rightleftharpoons (NH_2)_2CO(g) + H_2O(g)$$

A chemical plant produces 530 tonnes of urea per day. Calculate the theoretical mass, in tonnes, of ammonia required to produce 530 tonnes of urea.

  b) Add a line to a copy of the flow chart to show how the Bosch–Meiser process can be made more economical.

3 State three ways in which chemists can design an industrial process to minimise its impact on the environment.

4 Explain how each of the following factors could influence the design of an industrial process:

  a) energy requirements

  b) recycling.

Chapter 15

Chemical calculations

What you should know

- ★ Chemical equations, using formulae and **state symbols**, can be written and balanced.
- ★ The mass of a mole of any substance, in grams (g), is equal to the gram formula mass (GFM).
- ★ Calculations can be performed using the relationship between the mass and the number of moles of a substance.
- ★ For solutions, the mass of solute (grams or g), the number of moles of solute (moles or mol), the volume of solution (litres or l), or the concentration of the solution (moles per litre or mol 1^{-1}), can be calculated from data provided.
- ★ The molar volume (litres mol^{-1}) is the volume occupied by one mole of any gas at a certain temperature and pressure. The molar volume is the same for all gases at the same temperature and pressure.
- ★ Calculations can be performed using the relationship between the volume of gas, molar volume and the number of moles of a substance.
- ★ Calculations can be performed given a balanced equation using data including: GFM, masses, numbers of moles, concentrations and/or volumes of solutions, molar volumes, volumes for gases and percentage yield.
- ★ The efficiency with which reactants are converted into the desired product is measured in terms of the percentage yield and atom economy.
- ★ By considering a balanced equation, the limiting reactant and the reactant(s) in excess can be identified by calculation.
- ★ In order to ensure that a costly reactant is converted into product, an excess of the less expensive reactant(s) can be used.
- ★ The 'theoretical yield' is the quantity of desired product obtained, assuming full conversion of the limiting reagent, as calculated from the balanced equation.
- ★ The 'actual yield' is the quantity of the desired product formed under the prevailing reaction conditions.
- ★ For a particular set of reaction conditions, the percentage yield provides a measure of the degree to which the limiting reagent is converted into the desired product.
- ★ The percentage yield can be calculated using the equation:

$$\% \text{ yield} = \frac{\text{actual yield}}{\text{theoretical yield}} \times 100$$

- ★ Given costs for the reactants, a percentage yield can be used to calculate the cost of reactant(s) required to produce a given mass of product.
- ★ The atom economy measures the proportion of the total mass of all starting materials converted into the desired product in the balanced equation.
- ★ The percentage atom economy can be calculated using the equation:

$$\% \text{ atom economy} = \frac{\text{mass of desired product}}{\text{total mass of reactants}} \times 100$$

- ★ Reactions that have a high percentage yield may have a low atom economy value if large quantities of by-products are formed.

Making a new product requires chemists to work with chemical equations to allow them to calculate how much product they are likely to make or how much reactant they will require. For Higher Chemistry you should be comfortable working with equations involving masses, volumes, **concentrations** and moles.

Simple chemical calculations

The following examples will remind you of the relationship between moles, mass, GFM, volume and concentration.

> *Remember*
>
> $$Mole = \frac{mass}{GFM} \ and$$
>
> $$Mole = concentration \times volume$$

Examples

1 **Calculate**
   a) the GFM of H_2O
   b) the number of moles present in 144 g of H_2O
   c) the mass of 20 moles of H_2O.

Solution

1 a) The GFM is obtained by adding the atomic masses for each atom.
       $H_2O = (2 \times 1) + 16 = $ **18**

   b) $Mole = \dfrac{mass}{GFM} = \dfrac{144}{18} = $ **8**

   c) $Mass = mol \times GFM = 20 \times 18 = $ **360 g**.

2 **Calculate**
   a) the number of moles present in 2 litres of sulfuric acid, concentration 0.1 mol 1^{-1}
   b) the volume of a solution of containing 5 moles with a concentration of 0.25 mol 1^{-1}
   c) the concentration of 8 litres containing 0.2 mol of dissolved solute
   d) the concentration of 100 cm^3 containing 0.5 mol of dissolved solute.

Solution

2 Note that volumes must be in litres. To convert cm^3 into litres, divide by 1000 since 1 litre = 1000 cm^3.

   a) Mole = concentration × volume (or mol = CV)
      $Mol = 0.1 \times 2 = $ **0.2**

   b) $Volume = \dfrac{mol}{C} = \dfrac{5}{0.25} = $ **20 litres**

   c) $C = \dfrac{mol}{vol} = \dfrac{0.2}{8} = $ **0.025 mol 1^{-1}**

   d) $C = \dfrac{mol}{vol} = \dfrac{0.5}{0.1} = $ **5 mol 1^{-1}**

Calculations from equations involving mass

Example

Calculate the mass of water produced when 320 g of methane is burned according to the following equation:

$CH_4 + 2O_2 \rightarrow CO_2 + 2H_2O$

When faced with a question like this, it can help to work through it using the following five steps:

Step 1: Write a balanced chemical equation (unless already given).

Step 2: Identify the two chemicals referred to in the question and write the mole ratio.

Step 3: Calculate the moles of the substance you have been given a mass for (in this instance, methane). Note that gfm stands for gram formula mass.

Step 4: Use the mole ratio to calculate the number of moles of the other substance (in this case, water).

Step 5: Calculate the mass.

Solution

Step 1: $CH_4 + 2O_2 \rightarrow CO_2 + 2H_2O$

Step 2: 1 mole $CH_4 \rightarrow 2$ moles H_2O

Step 3: Mole $= \frac{mass}{gfm} = \frac{320}{16} = 20$ moles

Step 4: From our mole ratio in Step 2, we know that
20 moles $CH_4 \rightarrow 40$ moles H_2O.

Step 5: Mass $=$ moles $\times$ gfm

$\qquad = 40 \times 18$

$\qquad = 720$ g

Therefore, **720 g** of water is produced when 320 g of methane is burned.

Calculations from equations involving volumes and concentrations of solutions

Example

Calculate the concentration of hydrochloric acid used if 20 cm^3 of the acid was neutralised by 10 cm^3 of 1 mol l^{-1} sodium hydroxide solution.

In a question involving volumes and concentrations, it can help to work through six useful steps.

Step 1: Write a balanced chemical equation (unless already given).

Step 2: Identify the two chemicals referred to in the question and write the mole ratio.

Step 3: Write down the volumes, in litres, and the concentrations under the reactants.

Step 4: Calculate the number of moles of the chemical with the most information. Use moles = concentration × volume.

Step 5: Use the mole ratio to work out the moles of the other reactant (in this case, 1 mole of acid reacts with 1 mole of alkali).

Step 6: Calculate the volume, concentration or mass of the chemical you are asked to find out.

Solution

Step 1: $HCl + NaOH \rightarrow H_2O + NaCl$

Step 2: 1 mole of HCl reacts with 1 mole of NaOH

Step 3: Volume of HCl = 20 cm^3 = 0.02 l; concentration = ?
      Volume of NaOH = 10 cm^3 = 0.01 l; concentration = 1 mol l^{-1}

Step 4: Moles of NaOH = concentration × volume = CV =
      1 × 0.01 = 0.01

Step 5: Since 1 mole of alkali reacts with 1 mole of acid, moles
      HCl = 0.01

Step 6: Concentration of HCl = $\dfrac{\text{moles}}{\text{volume}}$

$$= \dfrac{0.01}{0.02}$$

$$= 0.5 \text{ mol l}^{-1}$$

Therefore, hydrochloric acid of concentration **0.5 mol l^{-1}** is used.

Calculations from equations involving masses, volumes and concentrations

Calculate the mass of calcium carbonate required to react completely with 300 cm³ of 0.1 mol l⁻¹ hydrochloric acid.

Solution

$$CaCO_3 + 2HCl \rightarrow CaCl_2 + CO_2 + H_2O$$

1 mole   2 moles

Moles of acid = CV = $0.1 \times 0.3 = 0.03$

Moles of $CaCO_3 = \frac{0.03}{2}$

So number of moles of $CaCO_3 = 0.015$ moles

Mass $CaCO_3$ = **moles × gfm**

Mass $CaCO_3$ = **0.015 × 100**

Therefore, **1.5 g** of calcium carbonate is needed to react completely with the acid.

Calculations from equations involving excess reactant

The previous examples have only considered the effect of one reactant on the quantity of product obtained. In reality, you would have to consider the quantities of all the reactants and then determine how they would affect the product. This can be done by deciding which reactant is in excess. The reactant in excess will not control how much product is obtained since some of this reactant will remain unreacted (there is too much of it). For these calculations, you should focus your attention on the reactant which is not in excess since this will control how much product is obtained. The reactant that is not in excess is known as the **limiting reagent**.

15 g of calcium carbonate were reacted with 50 cm³ of 4 mol l⁻¹ hydrochloric acid.

a) **Show by calculation which reactant was present in excess.**

b) **Calculate the mass of carbon dioxide produced.**

Solution

$CaCO_3 + 2HCl \rightarrow CaCl_2 + CO_2 + H_2O$

1 mol   2 mol   1 mol

a) Number of moles of $CaCO_3 = \frac{mass}{gfm}$

$$= \frac{15}{100}$$

$$= 0.15 \text{ mol}$$

Number of moles of HCl = C × V

$$= 4 \times \frac{50}{1000}$$

$$= 0.2 \text{ mol}$$

According to the equation, 1 mol of $CaCO_3$ neutralises 2 mol of HCl.

Hence, 0.1 mol of $CaCO_3$ neutralises 0.2 mol of HCl.

Since there is more than 0.1 mol of $CaCO_3$ present, this reactant is in excess.

b) To calculate the mass of carbon dioxide produced we use the quantity of the limiting reagent i.e. the hydrochloric acid.

According to the equation, 2 mol of HCl produce 1 mol of CO_2.

Hence, 0.2 mol of HCl produce 0.1 mol of CO_2.

Mass of CO_2 = moles × gfm

$$= 0.1 \times 44$$

$$= 4.4 \text{ g}$$

Therefore, **4.4 g** of carbon dioxide is produced in this reaction.

Example

1.2 g of magnesium was added to 80 cm^3 of 2 mol l^{-1} hydrochloric acid.

$Mg + 2HCl \rightarrow MgCl_2 + H_2$

a) Show by calculation which reactant was in excess.

b) Calculate the mass of hydrogen produced.

Solution

a) Number of moles of $Mg = \frac{1.2}{24} = 0.05$ moles

Number of moles of HCl = 2 × 0.08 = 0.16

From the equation, 1 mole of Mg reacts with 2 moles of HCl.

In other words, 0.05 mol Mg will react with 0.1 mol HCl.

Therefore, the HCl is in excess.

b) 1 mol Mg produces 1 mol H_2 so 0.05 mol Mg → 0.05 mol H_2

Mass of H_2 = 0.05 × 2 = 0.1 g

Therefore, **0.1 g** of hydrogen is produced in this reaction.

Calculations involving gases

In order to tackle calculations involving gases, we must consider the **molar volume**. This concept states that the volume of 1 mole of any gas is the same, provided the temperature and pressure are constant. The relationship between molar volume, volume and number of moles is shown in Figure 15.1.

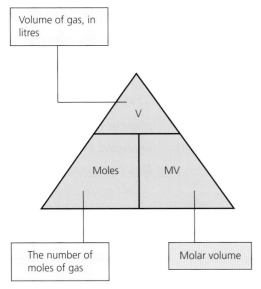

Figure 15.1 This triangle will help with calculations involving molar volume.

Example

The molar volume at 0 °C and 1 atmosphere pressure is 22.4 litres mol^{-1}. Calculate

a) the volume of 0.025 mol of oxygen

b) the number of moles of nitrogen in 4.48 litres under these conditions.

Solution

a) The volume of gas = moles × MV = 0.025 × 22.4 = **0.56 litres**

b) The number of moles of nitrogen = $\frac{V}{MV} = \frac{4.48}{22.4}$ = **0.2 moles**

Example

Assuming the molar volume is 24 litres mol^{-1}, calculate:

a) the volume of 3 mol of $SO_2(g)$

b) the number of moles of He in 200 cm^3 He gas.

Solution

a) Volume of SO_2 = 3 × 24 = **72 litres**

b) Number of moles = $\frac{0.2}{24}$ = **0.0083**

Example

Calculate the volume of carbon dioxide released when 0.4 g of calcium carbonate is dissolved in excess hydrochloric acid. The gas is collected at room temperature and pressure. The molar volume is 24 litres mol^{-1}. The equation for the reaction is

$$CaCO_3(s) + 2HCl(aq) \rightarrow CaCl_2(aq) + CO_2(g) + H_2O(l)$$

Solution

This calculation is treated in exactly the same way as previous calculations from equations except that we must consider the volume of gas rather than the mass.

1 mole of calcium carbonate $\rightarrow$ 1 mole of carbon dioxide gas

Number of moles of $CaCO_3 = \frac{0.4}{100}$

$= 0.004 \rightarrow 0.004$ moles of CO_2

1 mole of $CO_2 \rightarrow$ 24 litres

0.004 moles $\rightarrow 0.004 \times 24 = 0.096$ litres

Therefore, **0.096 litres** of carbon dioxide is released.

Hints & tips

When making a new product, chemists will usually react an excess of one chemical with another as this can improve the amount of product produced. When doing this, the excess reactant would normally be the one that is cheapest in order to maximise profits.

Comparing volumes of gases

Using the fact that the volume of 1 mole of a gas will be the same for all gases, at specific temperatures and pressures, we can carry out quick calculations for reactions where the reactants and products are gases. For example, nitrogen dioxide gas can be produced by reacting nitrogen with oxygen according to the equation:

$$N_2(g) + 2O_2(g) \rightarrow 2NO_2(g)$$

This balanced equation tells us that 1 mol of nitrogen reacts with 2 mol of oxygen to produce 2 mol of nitrogen dioxide. Comparing volumes, we could say that:

- 10 litres of N_2 would require 20 litres of O_2 to react and would produce 20 litres of NO_2
- 50 cm^3 of N_2 would require 100 cm^3 of O_2 to react and would produce 100 cm^3 of NO_2.

It is important to note that this concept is only applicable to gaseous reactants and products.

Example

Assuming that all volumes are measured at 150 °C and 1 atmosphere pressure, calculate

a) the volume of oxygen required for the complete combustion of 100 cm³ of methane

b) the volume of each product.

Solution

Table 15.1

	$CH_4(g) + 2O_2(g) \rightarrow CO_2(g) + 2H_2O(g)$			
Mole ratio	1	2	1	2
Volume ratio	1	2	1	2
Volumes given sin question	100 cm³			
Volume of oxygen required		200 cm³		
Volume of products			100 cm³	200 cm³

Hence 100 cm³ of methane

a) requires **200 cm³ of oxygen** for complete combustion

b) produces **100 cm³ of CO$_2$(g) and 200 cm³ of H$_2$O(g).**

Example

A mixture of 20 cm³ of propane and 130 cm³ of oxygen was ignited and allowed to cool. Calculate the volume and composition of the resulting gaseous mixture. All volumes are measured under the same conditions of room temperature and pressure.

Solution

Table 15.2

	$C_3H_8(g) + 5O_2(g) \rightarrow 3CO_2(g) + 4H_2O(l)$			
Mole ratio	1	5	3	4
Volume ratio	1	5	3	4
Volumes given in question	20 cm³	130 cm³		
Volume of oxygen required from reacting 20 cm³		100 cm³		
Volume remaining	0 cm³	30 cm³	60 cm³	

According to the equation,

20 cm³ of propane requires 5 × 20 cm³ of oxygen, i.e. 100 cm³.

Hence, oxygen is present in excess since its initial volume is 130 cm³.

Volume of excess oxygen = (130 − 100) cm³ = 30 cm³

Volume of carbon dioxide formed = (3 × 20) cm³ = 60 cm³

Therefore the resulting gas mixture consists of **30 cm³ of O$_2$** and **60 cm³ of CO$_2$.**

Example

Calculate the volume of carbon dioxide produced at room temperature and pressure when 0.4 g of calcium carbonate is added to 12 cm³ of 0.5 mol 1⁻¹ hydrochloric acid. The molar volume is 24 litres mol⁻¹.

Solution

$CaCO_3(s) + 2HCl(aq) \rightarrow CaCl_2(aq) + CO_2(g) + H_2O(l)$
1 mol   2 mol   1 mol
100 g   24 litres

Number of moles of calcium carbonate $= \dfrac{0.4}{100}$
$$= 0.004$$

Number of moles of hydrochloric acid $= 0.5 \times \dfrac{12}{1000}$
$$= 0.006$$

According to the equation, 1 mol of $CaCO_3$ requires 2 mol of HCl. Therefore, 0.004 mol $CaCO_3$ requires 0.008 mol HCl.

The number of moles of acid present is less than this, so the $CaCO_3$ is in excess and as a result the volume of CO_2 produced will depend on the number of moles of acid.

2 mol of HCl $\rightarrow$ 1 mol of CO_2
0.006 mol HCl $\rightarrow$ 0.003 mol CO_2

Hence, the volume of CO_2 produced $= 0.003 \times 24$ litres
$$= \textbf{0.072 litres (or 72 cm}^3\textbf{)}.$$

Percentage yield

The **percentage yield** is a simple way of comparing the amount of product actually obtained from a reaction with the amount expected. For example, if you calculated that you should obtain 400 kg of a medicine from a chemical reaction but when you did the reaction you only made 100 kg, the percentage yield would be 25%.

Remember

$$percentage\ yield = \frac{actual\ yield}{theoretical\ yield} \times 100$$

Example

12 g of CO_2 was obtained by reacting 6 g of C in excess oxygen. Calculate the percentage yield.

Solution

$C + O_2 \rightarrow CO_2$

Number of moles of $C = \frac{6}{12} = 0.5$

0.5 moles of $C \rightarrow 0.5$ moles CO_2

Mass of $CO_2 = 0.5 \times 44 = 22$ g

In other words, the theoretical yield of CO_2 is 22 g.

Percentage yield $= \frac{\text{actual yield}}{\text{theoretical yield}} \times 100$

$= \frac{12}{22} \times 100$

$= \mathbf{54.5\%}$

Example

4.5 g of H_2O was produced when 8 g of CH_4 was reacted with excess oxygen. Calculate the percentage yield.

Solution

$CH_4 + 2O_2 \rightarrow CO_2 + 2H_2O$

Number of moles of $CH_4 = \frac{8}{16} = 0.5$

0.5 moles $CH_4 \rightarrow 1$ mole H_2O

Mass of $H_2O = 1 \times 18 = 18$ g

In other words, the theoretical yield of H_2O is 18 g.

Percentage yield $= \frac{\text{actual yield}}{\text{theoretical yield}} \times 100$

$= \frac{4.5}{18} \times 100$

$= \mathbf{25\%}$

Example

Calculate the actual mass of SO_2 produced from reacting 28 g of sulfur with excess oxygen, according to the equation shown, assuming the percentage yield is 48%.

$S + O_2 \rightarrow SO_2$

Solution

Number of moles of $S = \frac{28}{32.1} = 0.872$

Since 1 mole of $S \rightarrow 1$ mole of SO_2, the mass of $SO_2 = 0.872 \times 64.1 = 55.9$ g

48% of 55.9 g $= \frac{48}{100} \times 55.9 = \mathbf{26.8\ g}$

Example

The salt lithium chloride can be obtained by reacting lithium oxide with excess hydrochloric acid.

$$Li_2O + 2HCl \rightarrow 2LiCl + H_2O$$

Calculate the cost of producing 10 kg of lithium chloride given the following information:

Table 15.3

Percentage yield of reaction	78%
Cost per kg of lithium oxide	£350

Solution

There are two parts to solving such questions:

1 calculating the mass of lithium oxide required to make 10 kg of LiCl
2 converting the calculated mass of lithium oxide into a cost.

Part 1: Calculation from equation:

$$Li_2O + 2HCl \rightarrow 2LiCl + H_2O$$

$$\text{Number of moles of LiCl required} = \frac{10{,}000}{42.4} = 235.849$$

This is the number of moles that are *actually* required.

$$\% \text{ yield} = \frac{\text{actual}}{\text{theoretical}} \times 100$$

$$78 = \frac{235.849}{\text{theoretical}} \times 100$$

$$0.78 = \frac{235.849}{\text{theoretical}}$$

$$\text{Theoretical} = \frac{235.849}{0.78} = 302.37$$

2 mol of LiCl is formed from 1 mol of Li_2O.

302.37 LiCl will be formed from 151.19 mol of Li_2O.

Mass of Li_2O = mol × GFM = 151.19 × 29.8 = 4505.32 g = **4.5 kg**

Part 2: Calculating the cost:

1 kg = £350

4.5 kg = 4.5 × 350 = **£1575**

Hints & tips ★

Be careful with rounding when performing calculations that involve multiple steps. Best practice is to use unrounded answers at each step and only round at the last step.

Using the percentage yield

Percentage yield is one of the factors taken into consideration when deciding on the best route to use to make a product. A reaction with a high percentage yield is always desired but will be rejected if the cost of reactants is very expensive. If an alternative reaction can be found with cheaper reactants, this might be used even if the percentage yield is lower. A cost analysis can be done to compare the costs for producing the same quantity of product.

Atom economy

The percentage yield tells us how successful the reaction is at converting reactants into products, but it does not give us information on how many by-products are formed. A reaction which produces lots of products can be problematic and wasteful. A reaction where most of the reactant atoms end up in the product is desirable. Chemists can apply the concept of **atom economy** to examine the proportion of reactants converted into the desired product.

Remember

$$\% \text{ atom economy} = \frac{\text{mass of desired product(s)}}{\text{total mass of reactants}} \times 100$$

Example

Hydrogen gas can be obtained by reacting methane gas with steam.

$CH_4 + H_2O \rightarrow CO + 3H_2$

Calculate the % atom economy for this reaction where hydrogen is the desired product.

Solution

Mass of desired product (from the equation) $= 3 \times$ gfm $H_2 = 3 \times 2 = 6$ g

Total mass of reactants $=$ gfm $CH_4 +$ gfm $H_2O = 16 + 18 = 34$ g

% Atom economy $= \frac{6}{34} \times 100 =$ **17.65%**

Example

Calculate the % atom economy for the production of propyl ethanoate, assuming that all reactants are converted into products, according to the following equation:

$C_3H_7OH + CH_3COOH \rightarrow C_3H_7OOCCH_3 + H_2O$

Solution

Mass of desired product (from the equation) $= 102$ g

Total mass of reactants $= 120$ g

% Atom economy $= \frac{102}{120} \times 100 =$ **85%**

This is a high atom economy which suggests that making the ester by this method does not create much waste. An examination of the equation shows that the other product is water which is easy to deal with as it is not toxic, flammable or highly corrosive.

Hints & tips

This chapter has examined several calculation types you are likely to encounter in Higher Chemistry. Chemists are expected to be numerate so you can expect several of these calculations in your final exam. Practise each calculation several times and develop a technique that works for you. Make sure you can recognise the different calculation types and apply the correct method. Layout of calculations is as important as accuracy. Take your time calculating formula masses, show your working and always go back and check your answers.

Key points

* The mass or volume of a product can be calculated using appropriate data and balanced chemical equations.
* Balanced chemical equations and calculations involving mass, volume and concentration can be used to calculate reactants in excess.
* The percentage yield can be found using the expression

$$\text{Percentage yield} = \frac{\text{actual yield}}{\text{theoretical yield}} \times 100$$

* The % atom economy can be calculated using the expression

$$\% \text{ atom economy} = \frac{\text{mass of desired product(s)}}{\text{total mass of reactants}} \times 100$$

Study questions

Calculation involving mass

1 Calculate the mass of carbon dioxide produced when 6 g of propane reacts completely with oxygen in the following reaction:

$C_3H_8 + 5O_2 \rightarrow 3CO_2 + 4H_2O$

Calculations involving mass, concentration and volume

2 Calculate the volume of $2 \, \text{mol} \, l^{-1}$ hydrochloric acid required to neutralise $100 \, \text{cm}^3$ of $1 \, \text{mol} \, l^{-1}$ sodium hydroxide.

$HCl + NaOH \rightarrow NaCl + H_2O$

3 Calculate the mass of magnesium required to react completely with $150 \, \text{cm}^3$ of $0.5 \, \text{mol} \, l^{-1}$ hydrochloric acid.

$Mg + 2HCl \rightarrow MgCl_2 + H_2$

Calculations involving excess reagent

4 Calculate the mass of water produced by reacting 8 g of hydrogen with 16 g of oxygen.

$H_2 + \frac{1}{2}O_2 \rightarrow H_2O$

5 Calculate the mass of hydrogen sulfide produced by reacting 24 g of iron (II) sulfide with $200 \, \text{cm}^3$ of $2 \, \text{mol} \, l^{-1}$ hydrochloric acid.

$FeS(s) + 2HCl(aq) \rightarrow FeCl_2(aq) + H_2S(g)$

6 In an experiment, 2 g of calcium oxide was reacted with $50 \, \text{cm}^3$ of $0.5 \, \text{mol} \, l^{-1}$ sulfuric acid.

$CaO + H_2SO_4 \rightarrow CaSO_4 + H_2O$

a) Show by calculation that the calcium oxide is in excess.

b) Calculate the mass of calcium sulfate produced.

Calculations involving gas volumes

(assume molar volume = $24 \, \text{litres} \, \text{mol}^{-1}$)

7 Calculate the volume of hydrogen produced when 6 g of magnesium is reacted with $100 \, \text{cm}^3$ of $1 \, \text{mol} \, l^{-1}$ hydrochloric acid.

$Mg + 2HCl \rightarrow MgCl_2 + H_2$

8 Copper carbonate decomposes to produce copper oxide and carbon dioxide. Calculate the volume of carbon dioxide produced from decomposing 40 g of copper carbonate.

$CuCO_3 \rightarrow CuO + CO_2$

Calculations involving gas volumes

9 State the volume and composition of remaining gases when $100\,cm^3$ propane is reacted with $600\,cm^3$ of oxygen gas.

$$C_3H_8(g) + 5O_2(g) \rightarrow 3CO_2(g) + 4H_2O(g)$$

10 Calculate the volume of hydrogen required to react completely with 4 litres of oxygen gas.

$$H_2 + \frac{1}{2}O_2 \rightarrow H_2O$$

Calculations involving percentage yield and atom economy

11 Ethyl 2-cyanoacrylate is synthesised from ethyl 2-cyanoethanoate by a process based on the Knovenagel reaction.

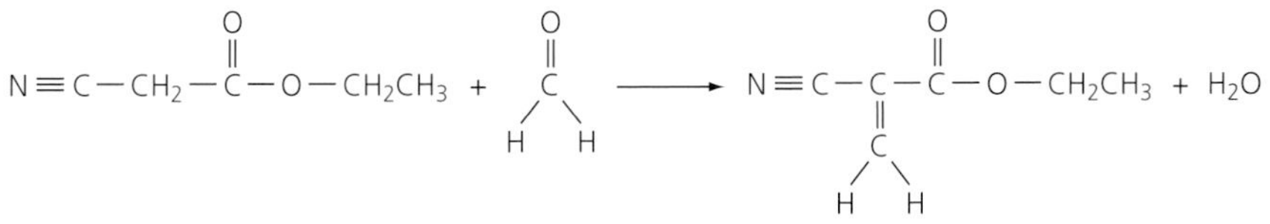

Ethyl 2-cyanoethanoate
mass of 1 mole = 113 g

Reactant A
mass of 1 mole = 30 g

Ethyl 2-cyanoacrylate
mass of 1 mole = 125 g

Water
mass of 1 mole = 18 g

Figure 15.2

a) Name reactant A.

b) Name this type of chemical reaction.

c) Calculate the % atom economy for the formation of ethyl 2-cyanoacrylate using this process. Show your working clearly.

d) In an experiment, 2 kg of ethyl 2-cyanoethanoate was reacted with an excess of reactant A to produce 1.5 kg of ethyl 2-cyanoacrylate. Calculate the percentage yield.

12 40 kg of ammonia, NH_3, was produced from 35 kg of hydrogen.

$$N_2 + 3H_2 \rightarrow 2NH_3$$

The percentage yield is

A 87.5%

B 43.75%

C 20.2%

D 38.1%

*13 Methanamide, $HCONH_2$, is widely used in industry to make nitrogen compounds. It is also used as a solvent as it can dissolve ionic compounds.

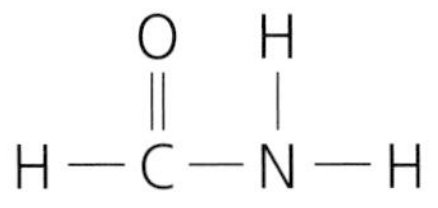

Figure 15.3

a) Why is methanamide a suitable solvent for ionic compounds?

b) In industry, methanamide is produced by the reaction of an ester with ammonia.

$$HCOOCH_3 \quad + \quad NH_3 \quad \rightarrow \quad HCONH_2 \quad + \quad CH_3OH$$

mass of 1 mole = 60.0 g   mass of 1 mole = 17.0 g   mass of 1 mole = 45.0 g   mass of 1 mole = 32.0 g

i) Name the ester used in the industrial production of methanamide.

ii) Calculate the % atom economy for the production of methanamide.

iii) 12 kg of ester was reacted with excess ammonia. Assuming an 82% yield, calculate the mass of methanamide actually produced.

Chapter 16
Controlling the rate

What you should know

★ Reaction rates must be controlled in industrial processes. If the rate is too low then the process will not be economically viable; if it is too high there will be a risk of explosion.

★ Calculations can be performed using the relationship between reaction time and relative rate with appropriate units.

★ Collision theory can be used to explain the effects of concentration, pressure, surface area (particle size), temperature and collision geometry.

★ A potential energy diagram can be used to show the energy pathway for a reaction.

★ Temperature is a measure of the average kinetic energy of the particles in a substance.

★ The activation energy is the minimum kinetic energy required by colliding particles before a reaction may occur.

★ Energy distribution diagrams can be used to explain the effect of changing temperature on the kinetic energy of particles and reaction rate.

★ The enthalpy change is the energy difference between the products and the reactants. The enthalpy change has a negative value for exothermic reactions or a positive value for endothermic reactions.

★ The activation energy is the minimum energy required by colliding particles to form an activated complex and can be calculated from potential energy diagrams. The activated complex is an unstable arrangement of atoms formed at the maximum of the potential energy barrier, during a reaction.

★ A catalyst provides an alternative reaction pathway with a lower activation energy.

★ A potential energy diagram can be used to show the effect of a catalyst on activation energy.

★ The effects of temperature and of adding a catalyst can be explained in terms of a change in the number of particles with energy greater than the activation energy.

The importance of reaction rate

Chemists carry out chemical reactions to make new products. It is desirable for products to be made quickly to maximise profit. However, there are many factors that affect the speed (rate) of a chemical reaction and so it is important that chemists are able to adjust these factors to control the reaction rate. If a reaction is too slow, it may not be economically viable, i.e. it may not generate a profit. On the other hand, if the reaction rate is too high, the reaction may go out of control and could explode!

Where it is difficult to measure a change in the chemical reaction, the time for the reaction is used to calculate the **relative rate** of reaction.

Measuring rate

The rate of a chemical reaction can be calculated by measuring changes in the concentration of reactants or products since reactants will always decrease and products will always increase. However, it can be difficult to measure this. An alternative is to measure the time taken for the reaction to reach a certain point, e.g. the time taken to produce a fixed volume of gas; the time taken for the reaction to lose a specific mass; or the time taken for a colour change to appear. This allows chemists to compare reactions since a reaction that takes a longer time is clearly slower than one that takes a shorter time.

The time recorded can be converted into a rate using the equation:

$$\text{rate} = \frac{1}{\text{time}}$$

For example, a reaction that took 20 s to reach completion would have a relative rate of:

$$\frac{1}{20} = 0.05 \text{ s}^{-1}$$

If you know the relative rate of a reaction, you can calculate the time taken for the reaction using the equation:

$$\text{time} = \frac{1}{\text{rate}}$$

As another example, the time taken for a reaction with a rate of 0.025s^{-1} would be:

$$\text{time} = \frac{1}{0.025} = 40 \text{ s}$$

Remember

$$rate = \frac{1}{time}$$

Example

Using the graph shown in Figure 16.1, calculate the time taken for the reaction when the temperature was 45 °C.

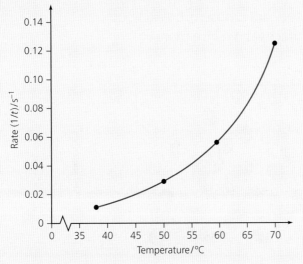

Figure 16.1 A graph of relative rate versus temperatre

Solution

At 45 °C, the relative rate is 0.02 s^{-1}.

$$\text{Time} = \frac{1}{\text{rate}} = \frac{1}{0.02} = \textbf{50 s}$$

Collision theory

For a successful chemical reaction to occur, reactant particles must collide. Some collisions result in a reaction, others do not. Successful collisions occur when

- the **collision geometry** is correct and
- the particles have the right amount of energy.

Concentration and pressure

Increasing the concentration, or pressure, increases the **rate of reaction** because you have more particles in the same space. As more of these particles are moving about, you are more likely to have collisions. If the particles colliding have sufficient energy, a successful reaction will occur.

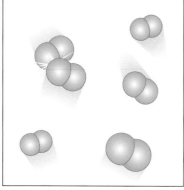
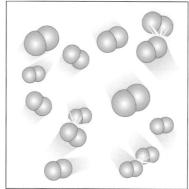

Lower concentration   Higher concentration

Figure 16.2 Particles are closer together if the concentration, or pressure, is increased. This leads to more collisions.

Particle size

Powdered solids react faster than lumps. Since it is only the particles on the surface of a solid that can react (since they are exposed), breaking up a solid into smaller pieces exposes more surfaces and hence more particles are available to react.

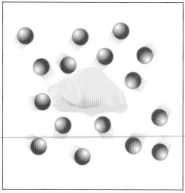

Smaller surface area   Larger surface area

Figure 16.3 Breaking up a solid into smaller particles increases the surface area leading to more collisions.

Temperature and activation energy

Raising the temperature of a reaction does more than simply increase the number of collisions between particles. Temperature is a measure of the average kinetic energy of the particles in a substance. If the temperature is increased, the particles have more kinetic energy. This means that they will collide with greater force.

It has been discovered that for a chemical reaction to occur, the colliding particles must have a minimum amount of kinetic energy. This minimum amount of energy is known as the **activation energy**. This concept helps to explain why some reactions do not occur at room temperature. For example, methane gas mixed with oxygen gas at room temperature does not react despite the particles colliding with each other. The particles do not have enough energy to react. To get the particles to react, some energy must be supplied (a spark or lit match will work).

In a gas, not all particles will have the same energy. *Energy distribution diagrams* can be used to show the energies of the particles. Figure 16.5 illustrates the distribution of kinetic energy and shows the minimum energy, activation energy, labelled E_A, required for a reaction to occur. The shaded area represents all of the **molecules** that have energy greater than the activation energy. In this example, very few particles have sufficient energy to react. This can be changed by giving the particles energy which can be done by increasing the temperature. Figure 16.6 shows that a small increase in temperature leads to a significant rise in the number of particles that have the minimum energy needed to react.

Figure 16.4 a) Collision is not successful because the particles do not have enough energy. b) Collision is successful as the particles do have enough energy.

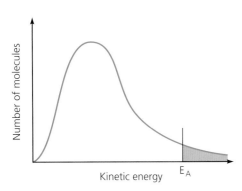

Figure 16.5 An energy distribution diagram showing that very few particles have enough energy to react.

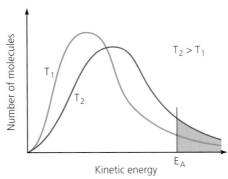

Figure 16.6 Increasing the temperature shifts the distribution curve to the right. Many more particles now have the minimum energy needed to react. The reaction rate increases.

Remember

- Temperature is a measure of the average kinetic energy of the particles in a substance.
- The activation energy is the minimum kinetic energy required by colliding particles before a reaction may occur.

Collision geometry

Collision geometry refers to the position of the reactants when they collide. Consider the reaction of propene with bromine as shown in Figure 16.7. Direct collision with the carbon to carbon double bond is more favourable and would be more likely to result in a reaction.

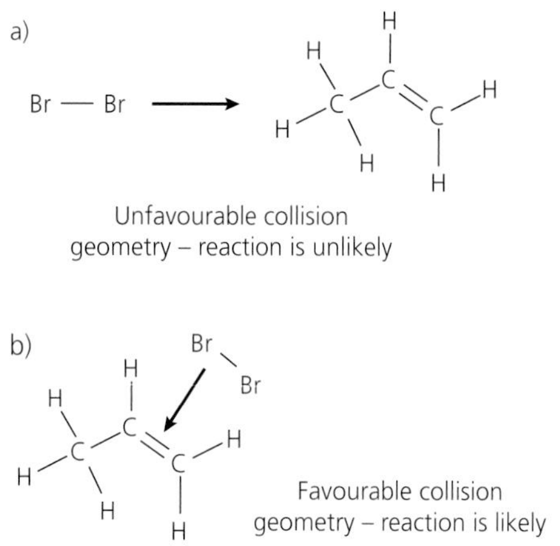

Hints & tips

Know and understand the distribution curve shown in Figure 16.5 and remember that increasing the temperature leads to an increase in reaction rate because more particles now have energy greater than or equal to the activation energy.

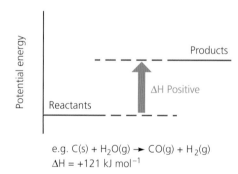

a)

Br — Br →

Unfavourable collision geometry – reaction is unlikely

b)

Favourable collision geometry – reaction is likely

Figure 16.7 The collision geometry must be correct for a collision to be successful.

Reaction profiles

The energy change that occurs when reactants are converted into products is known as the **enthalpy change** (ΔH) and can be shown on potential energy diagrams as in Figures 16.8 and 16.9.

In an **exothermic reaction** (Figure 16.8) the products have less energy than the reactants. Heat energy has been released to the surroundings. The ΔH for the reaction has a negative value to show that energy is 'lost' to the surroundings.

In an **endothermic reaction** (Figure 16.9) the products have more energy than the reactants as energy has been taken in from the surroundings. Removing heat energy from the surroundings causes the temperature of the surroundings to fall. Endothermic reactions have a positive ΔH to show that energy has been 'gained' from the surroundings.

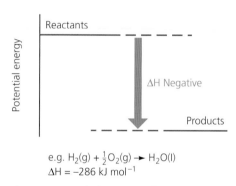

Reactants

Potential energy

ΔH Negative

Products

e.g. $H_2(g) + \frac{1}{2}O_2(g) \rightarrow H_2O(l)$
$\Delta H = -286$ kJ mol^{-1}

Figure 16.8 Enthalpy change for an exothermic reaction

Products

Potential energy

ΔH Positive

Reactants

e.g. $C(s) + H_2O(g) \rightarrow CO(g) + H_2(g)$
$\Delta H = +121$ kJ mol^{-1}

Figure 16.9 Enthalpy change for an endothermic reaction

Remember

$$\Delta H = H_{products} - H_{reactants}$$

Activation energy and the activated complex

Potential energy diagrams can be used to show the activation energy for a reaction, as shown in Figure 16.10.

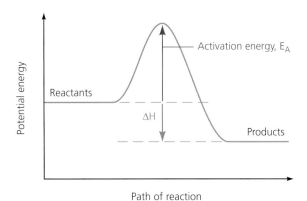

Figure 16.10 Potential energy diagram showing the activation energy and enthalpy change

As a reaction proceeds from reactants to products, a very-high-energy species known as an **activated complex** is formed. This is an unstable arrangement of atoms which is very high in energy. In potential energy diagrams, the activated complex is shown at the very top of the activation energy barrier as illustrated in Figure 16.11.

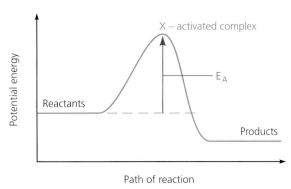

Figure 16.11 Activated complex

> *Remember*
>
> The activated complex is a high energy, unstable arrangement of atoms.

Catalysts

Catalysts speed up chemical reactions without being used up. Catalysts work by forming temporary bonds with reactants, causing the bonds within the reactants to weaken. This lowers the activation energy, allowing many more reactions to occur. The lowering of activation energy can be shown on a potential energy diagram as in Figure 16.12.

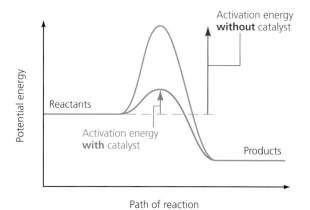

Figure 16.12 The lowering of the activation energy by a catalyst

The effect of a catalyst on an energy distribution diagram can be represented as shown in Figure 16.13. This shows that, as the activation energy has been lowered, many more reactants now have the minimum energy required to react. The particles that are now able to react are those shaded in orange.

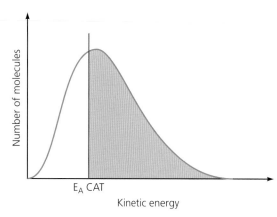

Figure 16.13 Effect of a catalyst on an energy distribution diagram

Hints & tips ⭐

Note that the effect of a catalyst on the energy distribution diagram is different from the effect of increasing temperature.

When the temperature is increased, the activation energy does not change. The rate increases as particles have more energy. When a catalyst is used, the activation does change: it is decreased.

Example 🚩

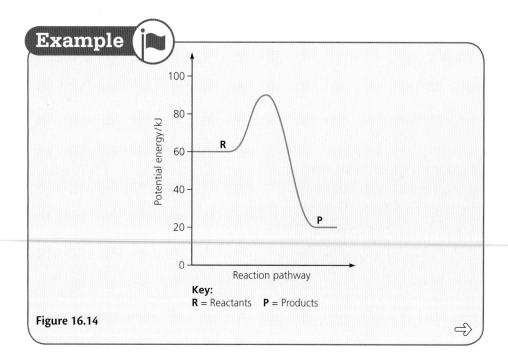

Key:
R = Reactants   **P** = Products

Figure 16.14

Using the potential energy diagram in Figure 16.14, calculate:

a) ΔH for the forward reaction

b) ΔH for the reverse reaction

c) E_A for the forward reaction

d) E_A for the reverse reaction.

e) **Suggest a value for the activation energy and enthalpy change for the forward reaction if a catalyst was used.**

f) **Suggest an energy value for the activated complex.**

Solution

a) $ΔH = H_{products} - H_{reactants} = 20 - 60 = $ **−40 kJ**

b) $ΔH = H_{products} - H_{reactants} = 60 - 20 = $ **40 kJ**

c) $E_A = 90 - 60 = $ **30 kJ**

d) $E_A = 90 - 20 = $ **70 kJ**

e) **Any value lower than 30 kJ for E_A.** The ΔH would still be 40 kJ as catalysts do not affect the enthalpy change.

f) **90 kJ**

Reaction rate graphs

Look at the graph shown in Figure 16.15.

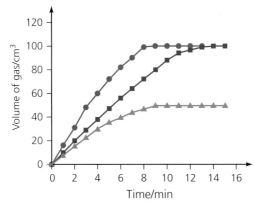

Figure 16.15 A graph showing rates of reaction for three versions of an experiment

The red line shows the results of an experiment where the volume of carbon dioxide released from the reaction of 1 g of $CaCO_3$ lumps with 1 mol l^{-1} excess hydrochloric acid was measured.

The blue line is steeper telling us that the reaction rate was higher; however, the final volume of gas released was the same in both instances: 100 cm^3. As the acid is in excess, the final volume of CO_2 is controlled by the mass of $CaCO_3$. $CaCO_3$ is the limiting reagent. Provided the mass of $CaCO_3$ stays at 1 g, the final volume of gas should stay the same.

The blue line could represent the reaction between:

● 1 g of $CaCO_3$ *powder* with 1 mol l^{-1} excess hydrochloric acid

● 1 g of $CaCO_3$ lumps with a *higher concentration* of excess hydrochloric acid

- 1 g of $CaCO_3$ lumps with 1 mol l^{-1} excess hydrochloric acid at a *higher temperature*
- 1 g of $CaCO_3$ lumps with 1 mol l^{-1} excess hydrochloric acid, using a *catalyst*.

The green line shows a final volume of gas which is half the original. We know that the volume of gas is controlled by the mass of $CaCO_3$, so the green line would be for the reaction of half the mass of $CaCO_3$ lumps with 1 mol l^{-1} excess hydrochloric acid, in other words, 0.5 g.

Key points

* Relative rate can be calculated using the equation:
$$\text{rate} = \frac{1}{\text{time}}$$

* Collision theory states that reactants must collide with the correct geometry and possess a minimum energy before a successful reaction occurs.
* Temperature is a measure of the average kinetic energy of the particles in a substance.
* The activation energy is the minimum kinetic energy required by colliding particles before a reaction may occur.
* Energy distribution diagrams help to explain why increasing the temperature significantly increases the reaction rate.
* Reaction profiles show the enthalpy change that occurs in a reaction and the activation energy.
* Exothermic reactions have a $-\Delta H$. Endothermic reactions have a $+\Delta H$.
* The activated complex is a high-energy, unstable arrangement of atoms. It is formed at the top of the activation energy barrier.
* Catalysts speed up a reaction by lowering the activation energy.

Study questions

1 Complete the following table to calculate the rate or time, including the correct units.

Table 16.1

Rate	Time
(a)	25 s
(b)	4 min
0.04 s^{-1}	(c)
0.0024 min^{-1}	(d)

2 Give a definition for the following terms:
a) activation energy
b) activated complex.

3 Potential energy distribution diagrams can be used to explain reaction rates.
a) Draw an energy distribution diagram and use it to explain the effect of increasing the temperature on reaction rate.
b) Draw an energy distribution diagram and use it to explain the effect of adding a catalyst on reaction rate.

4 Zinc was added to 25 cm³ of hydrochloric acid, concentration
   2 mol l⁻¹. Which of the following measurements, taken at regular
   intervals and plotted against time, would give the graph shown in
   Figure 16.16? The reaction is exothermic.
   A temperature
   B volume of gas produced
   C **pH** of solution
   D mass of the beaker and contents

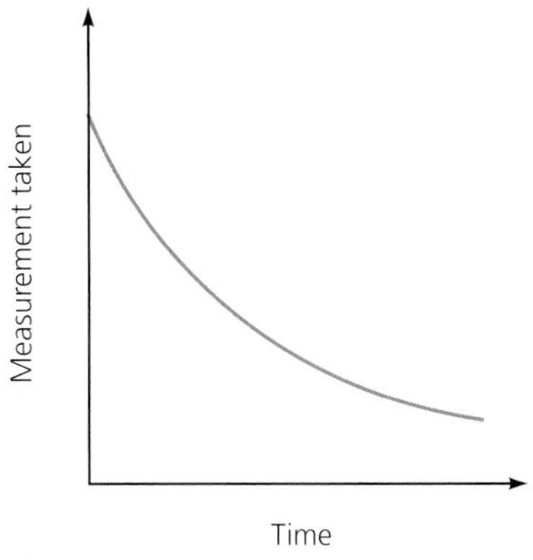

Figure 16.16

5 In which of the following will *both* changes result in an increase in the
   rate of a chemical reaction?
   A a decrease in activation energy and an increase in the frequency of
      collisions
   B an increase in activation energy and a decrease in particle size
   C an increase in temperature and an increase in the particle size
   D an increase in concentration and a decrease in the surface area of the reactant particles

6 Look at Figure 16.17.

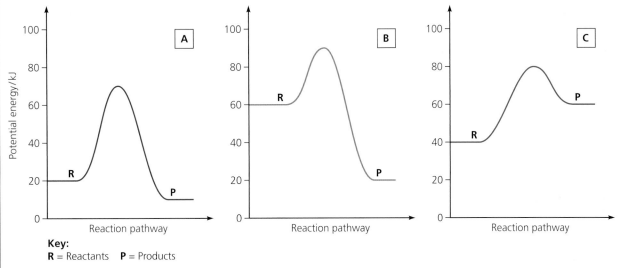

Key:
R = Reactants   **P** = Products

Figure 16.17

   a) Which reaction is likely to be the fastest?
   b) Which reaction is likely to have the most stable activated complex?
   c) Which reaction is endothermic?
   d) Which reaction has the greatest enthalpy change?
   e) Which reaction has an activation energy of 40 kJ?

7 Which of the following descriptions describes how a catalyst works in a chemical reaction?
   A It supplies energy to the reactants.
   B It does not take part in the reaction.
   C It lowers the energy required to form an activated complex.
   D It lowers the enthalpy change for the reaction.

8 Figure 16.18 shows the results of several experiments in which excess zinc metal was reacted with sulfuric acid. The red line represents the reaction of lumps of zinc metal with $100\,cm^3$ of $1.0\,mol\,l^{-1}$ sulfuric acid at $23\,°C$.

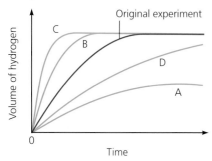

Figure 16.18

a) Draw a diagram of the apparatus that could be used to measure the volume of gas produced in this experiment.

b) Identify the line in the graph which shows the effect of
   i) increasing the temperature
   ii) increasing the temperature and adding a catalyst
   iii) using $100\,cm^3$ of $0.5\,mol\,l^{-1}$ sulfuric acid at $23\,°C$
   iv) using $100\,cm^3$ of $1.0\,mol\,l^{-1}$ sulfuric acid at $13\,°C$.

Chapter 17

Chemical energy

★ Enthalpy is a measure of the chemical energy in a substance.
★ A reaction that releases heat energy is described as 'exothermic'. In industry, exothermic reactions may require heat to be removed to prevent the temperature rising.
★ A reaction that takes in heat energy is described as 'endothermic'. In industry, endothermic reactions may incur costs in supplying heat energy in order to maintain the reaction rate.
★ The quantity of heat energy released can be determined experimentally and calculated using $E_h = cm\Delta T$.
★ The enthalpy of combustion of a substance is the enthalpy change when 1 mole of the substance burns completely in oxygen.
★ Hess's law states that the enthalpy change for a chemical reaction is independent of the route taken. The enthalpy change for a reaction can be calculated using Hess's law, given appropriate data.
★ The molar bond enthalpy is the energy required to break 1 mole of bonds in a diatomic molecule. A mean molar bond enthalpy is the average energy required to break 1 mole of bonds, for a bond that occurs in a number of compounds.
★ Bond enthalpies can be used to estimate the enthalpy change occurring for a gas phase reaction, by calculating the energy required to break bonds in the reactants and the energy released when new bonds are formed in the products.

Enthalpy

Enthalpy is a measure of the chemical energy in a substance. The change in energy in a chemical reaction is known as the enthalpy change. In the exothermic reaction illustrated by the energy profile diagram in Figure 17.1, the energy of the products is 20 kJ and the energy of the reactants is 60 kJ. The enthalpy change for this reaction is the difference between products and reactants, in other words $(20 - 60)kJ = -40\,kJ$. This tells us that the reaction is exothermic as 40 kJ has been 'lost' to the surroundings.

Chemists are interested in enthalpy changes, especially for industrial processes, since they can have health, safety and cost considerations. For example, a highly exothermic reaction could require specialist reaction vessels with cooling pipes to contain the reaction. On the other hand, a highly endothermic reaction may require expensive insulation and heating at a later stage, which can be very costly.

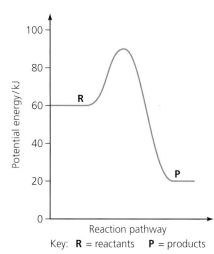

Key: **R** = reactants   **P** = products

Figure 17.1 The enthalpy change in a chemical reaction is the difference between the energy of the products and reactants.

Enthalpy of combustion

The **enthalpy of combustion** is defined as the enthalpy change when one mole of a substance is burned completely in oxygen. In the lab, the enthalpy of combustion of a fuel can be measured using a set up such as the one shown in Figure 17.2.

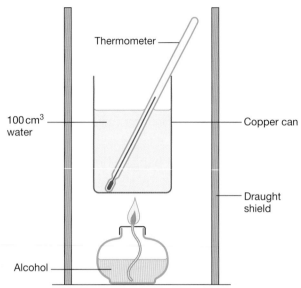

Thermometer

100 cm³ water

Copper can

Draught shield

Alcohol

Figure 17.2 This simple lab apparatus can be used to measure the enthalpy of combustion of a fuel.

Example

4.18 kJ of energy was released when 0.20 g of methanol (CH_3OH) was burned. Calculate the enthalpy of combustion.

Solution

To find the enthalpy of combustion you have to relate the energy released to the mass of 1 mole of methanol, 32 g.

0.20 g → 4.18 kJ

1.00 g → 20.9 kJ

32.00 g → 668.8 kJ

So, the enthalpy of combustion for methanol would be **−668.8 kJ mol⁻¹**.

Hints & tips

Remember, the enthalpy of combustion value will always have a negative value to show that the reaction is exothermic. In addition, the final answer which relates energy to 1 mole will have the units kJ mol⁻¹.

In the example, the energy released by burning methanol was given.

To calculate the energy released from the fuel, the following equation is used:

Remember

$$E_h = cm\Delta T$$

where

c is the specific heat capacity of water, $4.18 \, kJ \, kg^{-1} \, °C^{-1}$

m is the mass of water heated, in kg

ΔT is the change in temperature of the water.

Example

An experiment was carried out to calculate the enthalpy of combustion of ethanol. The results from the experiment are shown. Calculate the enthalpy of combustion of ethanol using this data.

Mass of ethanol burner at the start of the experiment = 40 g

Mass of ethanol burner at the end of the experiment = 39.8 g

Temperature of water at the start of the experiment = 22 °C

Temperature of water at the end of the experiment = 34 °C

Volume of water heated = 100 cm³

Solution

$E_h = cm\Delta T = 4.18 \times 0.1 \times 12 = 5.02 \, kJ$ (to two decimal places)

Mass of ethanol burned = 0.2 g

0.2 g $\rightarrow$ 5.02 kJ

1.0 g $\rightarrow$ 25.1 kJ

46 g $\rightarrow$ 1154.6 kJ

The enthalpy of combustion of ethanol is **−1155 kJ mol⁻¹**.

Hints & tips

*When using information like this, it is assumed that 1 cm³ of water weighs 1 g. Since the c value of 4.18 is the amount of energy required to raise the temperature of **1 kg** of water by 1°C, you should convert this mass of water into kg, i.e. 1 cm³ = 0.001 kg.*

Experimental concerns

When using a simple calorimeter such as the one shown in Figure 17.2, the values obtained for the enthalpy of combustion are usually much lower than the theoretical values. When carrying out these experiments, not all of the fuel combusts completely and there is always some heat lost to the surroundings. Any technique which addresses these experimental concerns will give improved enthalpy values.

Hess's law

In simple terms, **Hess's law** states the following:

Remember

The enthalpy change of a chemical reaction is independent of the route taken.

This allows chemists to calculate the enthalpy changes for reactions which are often difficult to carry out. When using Hess's law, it is usual to be presented with a direct route and several alternative routes. In most cases, the enthalpy change for the direct route is equal to the sum of the enthalpy changes for all other routes. This is illustrated by the examples which follow.

The first example illustrates how Hess's law can be used to calculate the enthalpy change when carbon is reacted with hydrogen to form methane, according to the equation:

$$C(s) + 2H_2(g) \rightarrow CH_4(g)$$

In this case, the direct route is the reaction of carbon with hydrogen to form methane. An alternative route must start with the same reactants (i.e. carbon and hydrogen), end with the same products (i.e. methane) and give us enthalpy information. Enthalpies of combustion can be used for carbon, hydrogen and methane since enthalpy data for combustion is easy to obtain.

Example

$$C(s) + 2H_2(g) \rightarrow CH_4(g)$$

Calculate the enthalpy change of the above reaction using the enthalpies of combustion of carbon, hydrogen and methane.

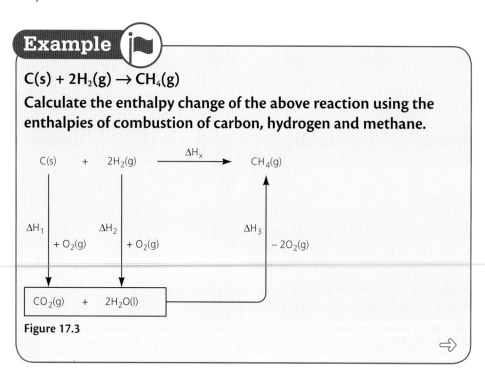

Figure 17.3

Solution

Step 1: Write equations for the enthalpy of combustion of C, H_2 and CH_4.

1 $C(s) + O_2(g) \rightarrow CO_2(g)$ $\qquad\qquad$ $\Delta H = -394 \, kJ \, mol^{-1}$

2 $H_2(g) + \frac{1}{2}O_2(g) \rightarrow H_2O(g)$ $\qquad\quad$ $\Delta H = -286 \, kJ \, mol^{-1}$

3 $CH_4(g) + 2O_2(g) \rightarrow CO_2(g) + 2H_2O(g)$ $\;\;$ $\Delta H = -891 \, kJ \, mol^{-1}$

The target equation is $C(s) + 2H_2(g) \rightarrow CH_4(g)$.

You have to use the ΔH combustion equations and rearrange them to resemble the target equation.

Step 2:

- Multiply equation 2) $\times$ 2 to give 2 moles of H_2.
- Reverse equation 3) so that CH_4 is a product. (If you reverse the equation, you must reverse the ΔH.)
- Rewrite all three equations and add to give the target.

1 **$C(s)$** $+ O_2(g) \rightarrow CO_2(g)$ $\qquad\qquad$ $\Delta H = -394 \, kJ \, mol^{-1}$

2 **$2H_2(g)$** $+ O_2(g) \rightarrow 2H_2O(g)$ $\qquad\quad$ $\Delta H = 2 \times -286 \, kJ \, mol^{-1}$

3 $CO_2(g) + 2H_2O(g) \rightarrow$ **$CH_4(g)$** $+ 2O_2(g)$ $\quad$ $\Delta H = +891 \, kJ \, mol^{-1}$

$\rule{10cm}{0.4pt}$

$\quad C(s) + 2H_2(g) \rightarrow CH_4(g)$ $\qquad\qquad\quad$ $\Delta H = -394 + -572 + 891$

$\qquad\qquad\qquad\qquad\qquad\qquad\qquad\qquad\qquad$ $= \mathbf{-75 \, kJ \, mol^{-1}}$

Example

The following equation shows the formation of ethanol from carbon, hydrogen and oxygen.

$$2C(s) + 3H_2(g) + \frac{1}{2}O_2(g) \rightarrow C_2H_5OH(l)$$

Use the enthalpies of combustion of carbon, hydrogen and ethanol to calculate the enthalpy change of this reaction.

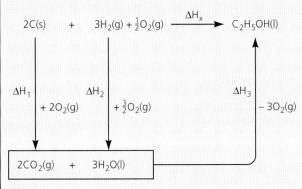

Figure 17.4

Solution

Step 1: Write equations for the enthalpy of combustion of C, H_2 and C_2H_5OH.

1 $C(s) + O_2(g) \rightarrow CO_2(g)$ $\qquad$ $\Delta H = -394\,kJ\,mol^{-1}$

2 $H_2(g) + \frac{1}{2}O_2(g) \rightarrow H_2O(g)$ $\qquad$ $\Delta H = -286\,kJ\,mol^{-1}$

3 $C_2H_5OH(l) + 3O_2(g) \rightarrow 2CO_2(g) + 3H_2O(l)$ $\quad$ $\Delta H = -1367\,kJ\,mol^{-1}$

The target equation is $2C(s) + 3H_2(g) \rightarrow \frac{1}{2}O_2(g) \rightarrow C_2H_5OH(l)$.

You have to rearrange the ΔH combustion equations to resemble this target equation.

Step 2:

- Multiply equation 1) × 2 to give 2 moles of C.
- Multiply equation 2) × 3 to give 3 moles of H_2.
- Reverse equation 3) so that C_2H_5OH is a product.
- Rewrite all three equations and add to give the target.

1 $2C(s) + 2O_2(g) \rightarrow 2CO_2(g)$ $\qquad$ $\Delta H = 2 \times -394\,kJ\,mol^{-1}$

2 $3H_2(g) + 1\frac{1}{2}O_2(g) \rightarrow 3H_2O(g)$ $\qquad$ $\Delta H = 3 \times -286\,kJ\,mol^{-1}$

3 $2CO_2(g) + 3H_2O(l) \rightarrow C_2H_5OH(l) + 3O_2(g)$ $\quad$ $\Delta H = +1367\,kJ\,mol^{-1}$

$$2C(s) + 3H_2(g) + \frac{1}{2}O_2(g) \rightarrow C_2H_5OH(l) \qquad \Delta H = \mathbf{-279\,kJ\,mol^{-1}}$$

Note: Oxygen is one of the elements present in ethanol but it is not involved in deriving the required enthalpy change. The calculation is based on enthalpies of combustion. Oxygen gas supports combustion; it does not have an enthalpy of combustion.

A quick method for solving Hess's law calculations

Example

$C(s) + 2H_2(g) \rightarrow CH_4(g)$

Calculate the enthalpy change for the above reaction using the enthalpies of combustion of carbon, hydrogen and methane.

Solution

Quick method

$\Delta H = 1 \times \Delta H_c + 2 \times \Delta H_{H_2} + (-\Delta H_{CH_4})$

$\quad = -394 + 2 \times (-286) + 891$

$\quad = \mathbf{-75\,kJ\,mol^{-1}}$

Example

$$2C(s) + 3H_2(g) + \tfrac{1}{2}O_2(g) \rightarrow C_2H_5OH(l)$$

Use the enthalpies of combustion of carbon, hydrogen and ethanol to calculate the enthalpy change for this reaction.

Solution

Quick method

$$\Delta H = 2 \times \Delta H_c + 3 \times \Delta H_{H_2} + (-\Delta H_{C_2H_5OH})$$

$$= 2 \times (-394) + 3 \times (-286) + 1367$$

$$= \mathbf{-279\ kJ\ mol^{-1}}$$

Hints & tips

In solving problems such as these, it is worth remembering the simple definition of Hess's law. Consider Figure 17.5, which illustrates the different routes which can be used to form water from hydrogen.

If you were asked to calculate an enthalpy change for reaction X, you would have to apply Hess's law. In such diagrams, always try to identify the direct route first. You can do this by looking for a start (where more than one arrow comes from) and an end (where more than one arrow points). Figure 17.5 tells us that $H_2(g)$ is the start and $H_2O\ (l)$ is the end. In other words, the direct route is $-286\ kJ\ mol^{-1}$. The sum of the enthalpy changes for all other routes must equal this value; therefore $-286 = -188 + X$. The value for X must be -98.

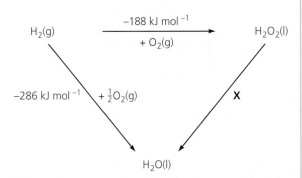

Figure 17.5

Example

Potassium chloride can be formed by the reaction of hydrochloric acid with solid potassium hydroxide (ΔH_1). An alternative route involves dissolving potassium hydroxide in water (ΔH_2) and then reacting the solution with hydrochloric acid (ΔH_3).

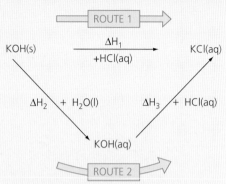

Figure 17.6 Making KCl(aq) via two routes

a) **Use Hess's law to write a statement which links the three enthalpy changes shown.**

b) **Calculate ΔH_2 given that $\Delta H_1 = -56\,kJ\,mol^{-1}$ and $\Delta H_3 = -37\,kJ\,mol^{-1}$.**

Solution

a) $\Delta H_1 = \Delta H_2 + \Delta H_3$

b) $\Delta H_2 = -19\,kJ\,mol^{-1}$

Example

*Some questions involving Hess's law present you with a 'target' equation and several other equations and ask you to form a relationship between them. This is illustrated in the following question, which is a past SQA exam question from 2009.

$$S(s) + H_2(g) \rightarrow H_2S(g) \qquad \Delta H = a$$
$$H_2(g) + \tfrac{1}{2}O_2(g) \rightarrow H_2O(l) \qquad \Delta H = b$$
$$S(s) + O_2 \rightarrow SO_2(g) \qquad \Delta H = c$$
$$H_2S(g) + 1\tfrac{1}{2}O_2(g) \rightarrow H_2O\,(l) + SO_2(g) \qquad \Delta H = d$$

What is the relationship between a, b, c and d?

A $a = b + c - d$
B $a = d - b - c$
C $a = b - c - d$
D $a = d + c - b$

Solution

The answers given tell you that the first equation **a** is the target; you are looking for equations which have S(s) and $H_2(g)$ as reactants and $H_2S(g)$ as a product. Applying this, you would require equation **c** for S(s), equation **b** for $H_2(g)$ and equation **d** for $H_2S(g)$. Equation **d** must, however, be reversed so that H_2S is a product. Reversing the equation reverses the ΔH sign. Overall, $a = c + b - d$ which is answer **A**.

Bond enthalpies

The data booklet lists *bond enthalpy* data for common molecules. This data can be used to calculate the enthalpy change for chemical reactions. When carrying out such calculations, it is worth remembering that *breaking bonds requires energy* while *making bonds releases energy*.

Example

Calculate the enthalpy change, using bond enthalpies, for the following reaction:

$H_2(g) + Cl_2(g) \rightarrow 2HCl(g)$

The necessary data are given in Tables 17.1 and 17.2.

Table 17.1

Bonds broken	$\Delta H/kJ\,mol^{-1}$
H–H	436
Cl–Cl	243
Total	**679**

Table 17.2

Bonds made	$\Delta H/kJ\,mol^{-1}$
H–Cl	−432
H–Cl	−432
Total	**−864**

Solution

Enthalpy change = total bonds broken + total bonds made

$\Delta H = 679 + (-864) = \textbf{−185 kJ mol}^{-1}$

Remember

The **molar bond enthalpy** is the energy required to break one mole of bonds in a diatomic molecule. This means it has a positive value. Making bonds will release energy so will have a negative value.

Key points

* The enthalpy change can be calculated for a chemical reaction using the equation $E_h = cm\Delta T$.
* Hess's law can be used to calculate the enthalpy change for a chemical reaction.
* Bond enthalpy data can be used to calculate the enthalpy change for a chemical reaction.

Study questions

1 0.05 mol of methane released 45 kJ of energy when burned. Calculate the enthalpy of combustion of methane.

2 12 g of sulfur released 80 kJ of energy when burned. Calculate the enthalpy of combustion of sulfur.

3 An experiment was carried out to determine the energy released when ethanol is burned. Use the data shown to calculate an experimental value for the enthalpy of combustion of ethanol.

Table 17.3

Mass of ethanol burner at the start of the experiment	82 g
Mass of ethanol burner at the end of the experiment	81.4 g
Temperature of water at the start of the experiment	23.1 °C
Temperature of water at the end of the experiment	38.3 °C
Volume of water heated	100 cm^3

4 The equation for the enthalpy of formation of ethyne is shown below. Using the enthalpies of combustion of carbon, hydrogen and ethyne, calculate the enthalpy of formation of ethyne.

$$2C(s) + H_2(g) \rightarrow C_2H_2(g)$$

5 Ethene gas can react with hydrogen gas to form ethane by the equation below. Using bond enthalpies, calculate the enthalpy change for this reaction.

$$C_2H_4(g) + H_2(g) \rightarrow C_2H_6(g)$$

*6 The equation for the combustion of diborane is shown below.

$$B_2H_6(g) + 3O_2(g) \rightarrow B_2O_3(s) + 3H_2O(l)$$

Calculate the enthalpy of combustion of diborane (B_2H_6) using the following data:

$$2B(s) + 3H_2(g) \rightarrow B_2H_6(g) \qquad \Delta H = 36 \text{ kJ mol}^{-1}$$

$$H_2(g) + \frac{1}{2}O_2(g) \rightarrow H_2O(l) \qquad \Delta H = -286 \text{ kJ mol}^{-1}$$

$$2B(s) + 1\frac{1}{2}O_2(g) \rightarrow B_2O_3(s) \qquad \Delta H = -1274 \text{ kJ mol}^{-1}$$

7 When chlorine gas reacts with methane in the presence of light, a free radical reaction occurs forming chloromethane and hydrogen chloride.
   a) Using bond enthalpies, calculate the enthalpy change for this reaction.
   b) Showing appropriate symbols and formulae, write an equation for the initiation step of this reaction.

Chapter 18

Equilibria

★ In a closed system, reversible reactions attain a state of dynamic equilibrium when the rates of forward and reverse reactions are equal.

★ At equilibrium, the concentrations of reactants and products remain constant, but are rarely equal.

★ To maximise profits, chemists employ strategies to move the position of equilibrium in favour of the products.

★ For a given reversible reaction, the effect of altering temperature or pressure or of adding/removing reactants/products can be predicted.

★ The addition of a catalyst increases the rates of the forward and reverse reactions equally.

★ The catalyst increases the rate at which equilibrium is achieved but does not affect the position of equilibrium.

The importance of equilibrium

The concept of **equilibrium** is applied to **reversible reactions** in a **closed system**. That is, where reactants and products are not allowed to escape. We say that a chemical reaction is in a state of equilibrium when the rate of the forward reaction is equal to the rate of the reverse reaction. Chemists try to influence chemical reactions by favouring the forward reaction – in other words, the making of the product – as this will maximise profit. This can be done by changing the concentration of reactants or products, changing the pressure and changing the temperature. If a reaction at equilibrium is subjected to a change in one of these factors, the equilibrium position will adjust to counteract this change, for example, if a reaction at equilibrium is cooled, the reaction will shift to counteract this change, i.e. the reaction which produces heat will be favoured.

Remember

A reaction is at equilibrium when the rate of the forward reaction = the rate of the reverse reaction.

Changing the concentration

Consider the reaction:

$$A + B \rightleftharpoons C + D$$

If the concentration of A or B is increased, the original equilibrium will be upset. As a result, more C and D will be produced until a new equilibrium is established. This is known as the equilibrium *shifting to the right*.

Likewise, if C or D was removed as it was produced, the equilibrium would again shift to the right, making more C and D.

This is illustrated by considering the equilibrium that exists in a bottle of bromine water:

$$Br_2(l) + H_2O(l) \rightleftharpoons Br^-(aq) + BrO^-(aq) + 2H^+(aq)$$

When the equilibrium shifts, there is a detectable colour change as the bromine liquid has a brown colour and the product ions are colourless.

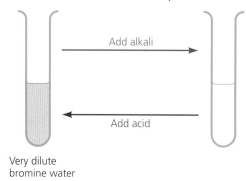

Very dilute
bromine water

Figure 18.1 Changing the position of equilibrium by altering the concentration of reactants or products

Adding $OH^-(aq)$, from an alkali such as NaOH(aq), removes H^+ ions to form water. This causes the equilibrium to shift to the right which results in the solution becoming colourless. If an acid is added (a source of H^+ ions) the equilibrium will shift to the left to counteract the change. This causes the solution to adopt a brown colour.

Example 🚩

Iodine monochloride (a brown liquid) can react with chlorine gas to form the yellow solid iodine trichloride:

$$ICl(l) + Cl_2(g) \rightleftharpoons ICl_3(s)$$

State what would be observed if the following changes were made to this reaction at equilibrium:

a) excess chlorine was added
b) chlorine gas was removed from the reaction.

Solution

a) Adding more chlorine would cause the equilibrium to shift to the right, producing more yellow $ICl_3(s)$.
b) Removing the chlorine would cause the equilibrium to shift to the left, causing the yellow solid to be converted into chlorine gas and brown liquid ICl.

Changing the temperature

In a reversible reaction, if the forward reaction is exothermic, the reverse reaction must be endothermic. If a reaction at equilibrium is exposed to a rise in temperature, the equilibrium will shift to favour the side which absorbs heat. In other words, the reaction will shift to the endothermic side. A fall in temperature will cause the reaction to favour the exothermic process. This is illustrated by observing the equilibrium that exists between dinitrogen tetroxide (a colourless gas) and nitrogen dioxide (a brown gas):

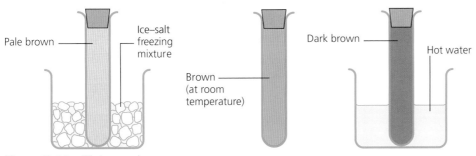

$$N_2O_4(g) \rightleftharpoons 2NO_2(g) \quad \Delta H = +ve$$
colourless   brown

Pale brown —— | Ice–salt freezing mixture

Brown (at room temperature) ——

Dark brown —— | Hot water

Figure 18.2 Equilibrium and temperature

Increasing the temperature causes the equilibrium to shift to the right (the endothermic reaction) which results in the gas mixture darkening in colour.

Lowering the temperature causes the equilibrium to shift to the left (the exothermic reaction) which results in the gas mixture lightening in colour.

Example

Steam reforming of coal can be used to produce a valuable mixture of carbon monoxide and hydrogen. The forward reaction is endothermic. State how a change in temperature on this reaction at equilibrium could affect the yield of CO and H_2.

$$C + H_2O \rightleftharpoons CO + H_2 \qquad \Delta H = +131\,kJ\,mol^{-1}$$

Solution

Increasing the temperature will increase the yield of CO and H_2 since the forward reaction is endothermic.

Hints & tips ⭐

When thinking about the effect of temperature on equilibrium it can be useful to imagine a house with a central heating system and an air conditioning system controlled by a thermostat. The thermostat detects when the temperature changes and will adjust conditions in the house so that the temperature gets back to the desired level. If the temperature falls, for example, in the winter, a signal is sent to the boiler to increase the central heating output so that the house is heated. If the temperature rises beyond a comfortable level, such as in the summer, a signal is sent to turn on the air conditioning to cool the air back to the desired temperature. You can think of the boiler as the exothermic process and the air conditioning as the endothermic process.

A rise in temperature will favour the endothermic process. A fall in temperature will favour the exothermic process.

Changing the pressure

Changing the pressure of gases at equilibrium causes a shift to readjust the pressure. In other words, if the pressure is increased, the equilibrium will shift to the side which causes the pressure to decrease. If the pressure is decreased, the equilibrium will shift to the side which causes an increase in pressure. In chemical reactions, the side of the equation with the most gaseous atoms or molecules will have the highest pressure.

For example:

$$2SO_2(g) + O_2(g) \rightleftharpoons 2SO_3(g)$$

In this equation, the left-hand side has 3 moles of gas while the right-hand side has 2 moles of gas. Increasing the pressure will cause the equilibrium to shift to the right.

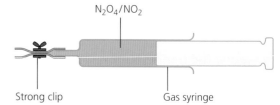

Consider the N_2O_4–NO_2 system which we met earlier in this chapter:

$$N_2O_4(g) \rightleftharpoons 2NO_2(g)$$
colourless   brown

Figure 18.3 Pushing in the plunger of this gas syringe causes the pressure to increase. A lightening of colour is observed as the equilibrium shifts to the left to produce more colourless N_2O_4 molecules.

Increasing the pressure will cause a shift to the left. Decreasing the pressure will cause a shift to the right.

Hints & tips ⭐

When considering the effect of pressure it is important to remember that only reactants and products in the gaseous state should be considered. Solids, liquids and solutions should be ignored. We can use the reaction between chlorine and water as an example.

$$Cl_2(g) + H_2O(l) \rightleftharpoons Cl^-(aq) + ClO^-(aq) + 2H^+(aq)$$

In this equilibrium, there is 1 mole of gas on the left-hand side and 0 moles on the right-hand side. Increasing the pressure would cause a shift to the right. Where there is an equal number of moles of gas on both sides, pressure will have no effect on the equilibrium position. For example:

$$CO(g) + H_2O(g) \rightleftharpoons CO_2(g) + H_2(g)$$

The effect of a catalyst

Catalysts have no effect on the position of equilibrium. As can be seen from Figure 18.4, catalysts work by lowering the activation energy. Since they lower the activation energy of both the forward and reverse reaction by the same amount, they do not affect the position of equilibrium. They do, however, allow the reaction to reach equilibrium more quickly.

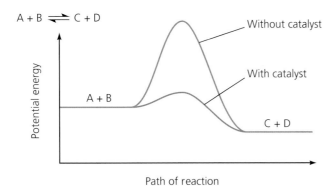

Figure 18.4 Potential energy diagram for a catalysed and uncatalysed reaction

Key points !

* A reversible reaction is said to have attained a state of dynamic equilibrium when the rate of the forward reaction is equal to the rate of the reverse reaction.
* Concentration of reactants or products, temperature and pressure can all affect a system at equilibrium. The system will adjust to counteract the change.
* A catalyst does not affect the position of equilibrium.

Summary

Table 18.1

Change applied	Effect on equilibrium position
Concentration Addition of reactant or removal of product Addition of product or removal of reactant	Equilibrium shifts to the right Equilibrium shifts to the left
Temperature Increase Decrease	Shifts in direction of endothermic reaction Shifts in direction of exothermic reaction
Pressure Increase Decrease	Shifts in direction which reduces the number of molecules in gas phase Shifts in direction which increases the number of molecules in gas phase
Catalyst	No effect on equilibrium position; equilibrium more rapidly attained

Study questions

1 Which of the following statements about the use of a catalyst in a reaction at equilibrium is false?
   A The rate of the reverse reaction will be increased.
   B The time taken to reach equilibrium will decrease.
   C The position of equilibrium will shift to the right.
   D The activation energy of the reaction will decrease.

2   In which of the following reactions will an increase in pressure have no effect on the position of equilibrium?

A $2SO_2(g) + O_2(g) \rightleftharpoons 2SO_3(g)$

B $N_2(g) + 2O_2(g) \rightleftharpoons 2NO_2(g)$

C $H_2(g) + Cl_2(g) \rightleftharpoons 2HCl(g)$

D $N_2 + 3H_2 \rightleftharpoons 2NH_3(g)$

3   The chromate/dichromate equilibrium is useful for studying factors that can affect equilibrium since the ions have a different colour.

$$2CrO_4^{2-}(aq) + 2H^+(aq) \rightleftharpoons Cr_2O_7^{2-}(aq) + H_2O(l)$$
   (yellow)   (orange)

a) Explain why adding KOH(aq) causes the solution to become more yellow in colour.

b) State the effect of adding sulfuric acid to the solution.

4   The industrial process used to manufacture ammonia is known as the **Haber process**. This involves reacting nitrogen with hydrogen to form ammonia:

$$N_2 + 3H_2 \rightleftharpoons 2NH_3(g) \quad \Delta H = -92 \, kJ$$

a) With reference to this equation, explain the effect of increasing the temperature on the
   i) speed of reaction
   ii) yield of ammonia.

b) With reference to this equation, explain the effect of increasing the pressure.

c) In the Haber process, the ammonia formed is continuously removed. State how this affects the position of equilibrium.

5   Phosphorus pentachloride decomposes according to the following equation:

$$PCl_5(g) \rightleftharpoons PCl_3(g) + Cl_2(g) \quad \Delta H = +124 \, kJ$$

State the effect on the concentration of PCl_3 if

a) the concentration of chlorine is increased

b) the PCl_3 formed is continuously removed

c) the temperature is decreased

d) the pressure is increased.

What you should know

★ Chromatography is a technique used to separate the components present within a mixture.
★ Chromatography separates substances by making use of differences in their polarity or molecular size.
★ The details of any specific chromatographic method or experiment are not required.
★ Depending on the type of chromatography used, the identity of a component can be indicated either by the distance it has travelled, or by the time it has taken to travel through the apparatus (retention time).
★ The results of a chromatography experiment can sometimes be presented graphically, showing an indication of the quantity of substance present on the y-axis and retention time of the x-axis.
★ Volumetric analysis involves using a solution of accurately known concentration in a quantitative reaction to determine the concentration of another substance.
★ Titration is used to determine, accurately, the volumes of solution required to reach the end-point of a chemical reaction. An indicator is normally used to show when the end-point is reached. Titre volumes within 0.2 cm^3 are considered concordant.
★ Solutions of accurately known concentration are known as standard solutions.
★ Redox titrations are based on redox reactions. In titrations using acidified permanganate, an indicator is not required, as purple permanganate solution turns colourless when reduced.
★ Given a balanced equation for the reaction occurring in any titration, the
  ★ concentration of one reactant can be calculated given the concentration of the other reactant and the volumes of both solutions
  ★ volume of one reactant can be calculated given the volume of the other reactant and the concentrations of both solutions.

Chromatography

Chromatography is used to separate the components in a mixture. Figure 19.1 shows an example of paper chromatography.

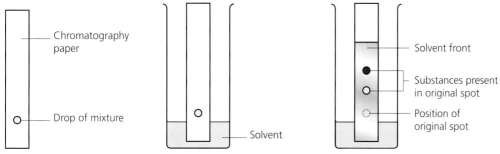

Figure 19.1 Paper chromatography

In the example shown in Figure 19.1, the original mixture separates into two components. Separation usually depends on
● the size of the molecules
● the polarity of the molecules.

A component can often be identified by how far it has travelled. For example, if the paper chromatography shown in Figure 19.1 was run using a non-polar solvent such as hexane, non-polar components would be attracted to the hexane and would be expected to travel further up the paper than polar components. In this case, it would be deduced that the red spot is less polar than the yellow spot.

In other forms of chromatography such as **gas liquid chromatography** (**GLC**), the results of the experiment are shown graphically as in Figure 19.2.

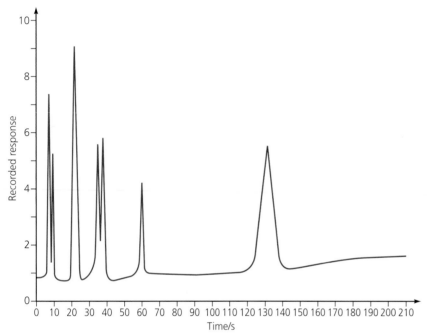

Figure 19.2 A graph from a GLC (chromatogram)

Fig 19.2 shows the **retention time** for each component: peaks with a short time have gone through the chromatography column quickly. Again, this is related to the size and/or the polarity of the components. For example, the peak at 130 s represents the component with the longest retention time, i.e. the component that has taken the longest time to travel through the column. This could be the biggest molecule.

Remember

Chromatography separates components based on molecular size or polarity.

Example

In an arson investigation, a chromatogram was obtained from a sample of fresh petrol and compared to a chromatogram obtained from a partially evaporated petrol sample from a piece of fabric.

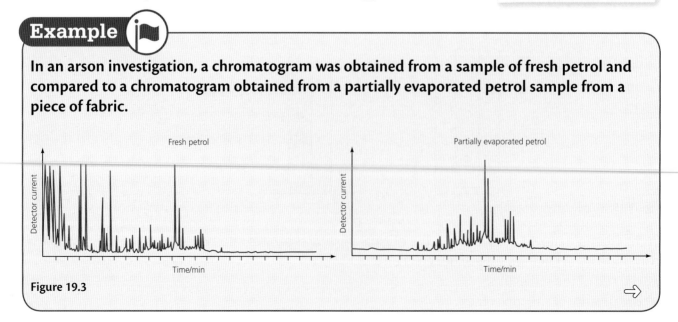

Figure 19.3

Explain how these chromatograms show that large molecules have longer retention times than small molecules in this type of chromatography.

Solution

Small molecules would be expected to evaporate more quickly than larger molecules. Since the chromatogram for the partially evaporated petrol shows an absence of peaks with short retention times, this would suggest that small molecules have a short retention time and, therefore, large molecules have a longer retention time.

Volumetric analysis

Volumetric analysis involves using a solution of known concentration to determine the concentration of an unknown solution. This is usually done using titrations which require the use of pipettes, burettes, **indicators** and standard flasks.

A solution of known concentration is known as a **standard solution**. The steps required to make a standard solution are shown in Figure 19.4.

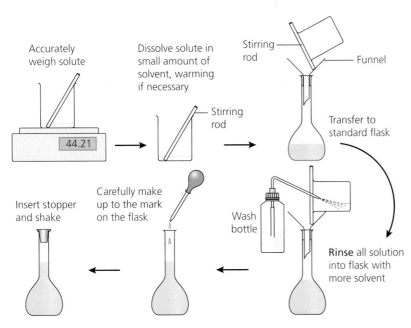

Figure 19.4 Making a standard solution

A common example of volumetric analysis is an acid–alkali titration. Consider an experiment where a known concentration of hydrochloric acid is titrated with an unknown concentration of sodium hydroxide solution.

By adding a fixed volume of the alkali into a flask, along with a suitable indicator, the volume of acid required to completely neutralise the alkali can be determined (Figure 19.5). The acid is added from the burette until the **end-point** is reached, as shown by the indicator changing colour.

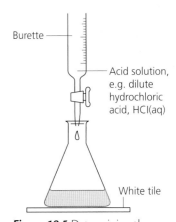

Figure 19.5 Determining the concentration of an alkali by titrating with an acid of known concentration

The volume of acid required to reach this end-point can be read from the burette. To improve accuracy, more than one titration is performed until concordant results are obtained (results within 0.2 cm³ of each other) as shown in the second worked example. Once this volume is known, the number of moles of acid can be calculated since its concentration is already known. Now, a balanced chemical equation can be used to find out the unknown concentration of the alkali. This is illustrated by the following two examples.

Example

25 cm³ of sodium hydroxide solution was added to a flask and titrated with 0.1 mol l⁻¹ hydrochloric acid. The volume of hydrochloric acid required to neutralise the sodium hydroxide was found to be 10.0 cm³. Use these results to calculate the concentration of the sodium hydroxide solution.

Solution

$HCl(aq) + NaOH(aq) \rightarrow NaCl(aq) + H_2O(l)$

Table 19.1

	HCl(aq)	NaOH(aq)
Mole ratio	1	1
Concentration, C	0.1 mol l⁻¹	?
Volume, V	10 cm³	25 cm³

From the information given in Table 19.1, the number of moles of hydrochloric acid reacting can be calculated using the equation moles = CV, where V is measured in litres.

Moles $= 0.1 \times 0.01 = 0.001$

From the mole ratio, 1 mole of HCl reacts with 1 mole of NaOH.

Thus, 0.001 moles of HCl would react with 0.001 moles of NaOH.

Concentration of NaOH $= \frac{\text{moles}}{\text{volume}} = \frac{0.001}{0.025} = \textbf{0.04 mol l⁻¹}$

Example

In a titration, it was found that 10 cm³ of potassium hydroxide was neutralised by 0.05 mol l⁻¹ sulfuric acid. The volumes of sulfuric acid used in the titration are recorded in Table 19.2.

Table 19.2

Titration	Volume of 0.05 mol l⁻¹ sulfuric acid/cm³
1	18.0
2	17.4
3	17.3

Calculate the concentration of the potassium hydroxide given that potassium hydroxide reacts with sulfuric acid according to the equation shown.

$$2KOH(aq) + H_2SO_4(aq) \rightarrow K_2SO_4(aq) + 2H_2O(l)$$

Solution

Average volume of sulfuric acid $= \frac{17.3 + 17.4}{2} = 17.35$ cm^3

(See Appendix 1 for more details on calculating the average and eliminating rogue data.)

Number of moles of sulfuric acid reacting $= CV = 0.05 \times 0.01735 = 0.00087$

According to the equation, 1 mole of H_2SO_4 reacts with 2 moles of KOH.

Thus, 0.00087 moles of H_2SO_4 will react with 0.0017 moles of KOH.

Concentration of KOH $= \frac{\text{moles}}{\text{volume}} = \frac{0.0017}{0.01} = \textbf{0.17 mol l}^{-1}$

Redox titrations

The concept of volumetric titrations can be applied to redox reactions. For example, a solution of potassium permanganate of known concentration can be used to determine the quantity of iron in iron tablets. The steps for this experiment are shown in Figure 19.6.

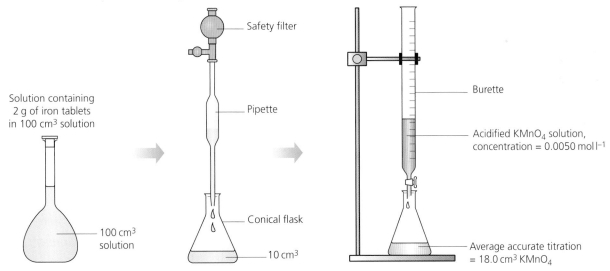

Figure 19.6 Determining the mass of iron in an iron tablet by volumetric titration

The redox reaction for this experiment is:

$$MnO_4^- + 8H^+ + 5e^- \rightarrow Mn^{2+} + 4H_2O$$
$$5Fe^{2+} \rightarrow 5Fe^{3+} + 5e^-$$

$$MnO_4^- + 8H^+ + 5Fe^{2+} \rightarrow Mn^{2+} + 4H_2O + 5Fe^{3+}$$

In other words, 1 mole of potassium permanganate will react with 5 moles of iron (II) ions.

From the experiment shown, the moles of permanganate required can be calculated:

Moles $= CV = 0.005 \times 0.018 = 9 \times 10^{-5}$ moles

Moles of $Fe^{2+} = 5 \times$ moles of permanganate $= 5 \times (9 \times 10^{-5}) = 0.00045$ moles

In other words, number of moles of Fe^{2+} present in $10\,cm^3$ solution $= 0.00045$ moles

Number of moles of Fe^{2+} present in $100\,cm^3$ solution $= 0.0045$ moles

Mass of iron present $=$ moles $\times$ gfm $= 0.0045 \times 55.8 = 0.25\,g$

Example

To determine the concentration of an iron (II) sulfate solution by titration with a potassium permanganate solution of known concentration

$20\,cm^3$ of iron (II) sulfate solution is transferred by pipette to a conical flask and excess dilute sulfuric acid is added. Potassium permanganate solution ($0.02\,mol\,l^{-1}$) is added from the burette until the contents of the flask just turn from colourless to purple, initial and final burette readings being noted. The titration is repeated to obtain concordant titres.

Since the permanganate solution is so strongly coloured compared to the other solutions, the reaction is self-indicating and the change at the end-point from colourless to purple is quite sharp.

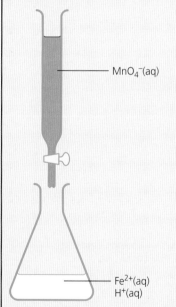

MnO₄⁻(aq)

Fe²⁺(aq)
H⁺(aq)

Figure 19.7 The redox titration between permanganate and iron (II) is an example of a self-indicating reaction since the purple colour of permanganate appears once all the iron (II) ions have reacted.

Calculate the concentration of an iron (II) sulfate solution given that 20.0 cm³ of it reacted with 24.0 cm³ of 0.02 mol l⁻¹ potassium permanganate solution. The redox equation for the reaction is:

$$MnO_4^- + 8H^+ + 5Fe^{2+} \rightarrow 5Fe^{3+} + Mn^{2+} + 4H_2O$$

Hint & tips

When solving any volumetric/redox titration questions, you should look for
a) a balanced chemical equation
b) information on concentration and volume.
You will always be able to calculate the number of moles of one substance using moles = CV. You can then use the mole ratio from the equation to calculate the number of moles of the 'unknown' substance. Dividing this by the volume shown will allow you to work out a concentration.

Solution

According to the equation, 1 mole of MnO_4^- oxidises 5 moles of Fe^{2+}.

Number of moles of MnO_4^- used $= CV = 0.02 \times 0.024 = 4.8 \times 10^{-4}$

Hence, number of moles of Fe^{2+} present $= 5 \times 4.8 \times 10^{-4} = 2.4 \times 10^{-3}$

This is contained in $20\,cm^3$, i.e. 0.02 litres.

Hence, concentration of $Fe^{2+}(aq)$, $C = \frac{2.4 \times 10^{-3}}{0.02} = 0.12$ mol l^{-1}

Since 1 mole $FeSO_4(aq)$ contains 1 mole of $Fe^{2+}(aq)$, the concentration of $FeSO_4(aq) = \textbf{0.12 mol l}^{-1}$.

Redox titrations, such as the permanganate/iron titration, are said to be self-indicating; there is no need for an indicator as the purple permanganate colour is used to judge the end-point. The solution will remain colourless provided there are iron (II) ions to react with the purple permanganate ions. Once all of the iron (II) ions have reacted, the end-point – the purple colour – will appear.

* Chromatography separates substances by making use of differences in their polarity or molecular size.
* Chromatograms are used to show how far a component has travelled or present graphical data showing retention time.
* A standard solution is a solution of accurately known concentration.
* The end-point of a titration is the point at which the reaction is just complete.
* An indicator is a substance which changes colour at the end-point.
* Some titrations are self-indicating.
* Redox titrations can be used to determine the concentration of a substance.

Study questions

1 The concentration of an iodine solution was determined by a redox titration. $10\,cm^3$ of a standard sodium sulfite solution, 0.1 mol l^{-1}, was transferred into a conical flask. It was found that $24.3\,cm^3$ of iodine was required to reach the end-point of the titration. The redox equation for the reaction is:

$$SO_3^{2-} + I_2 + H_2O \rightarrow 2I^- + SO_4^{2-} + 2H^+$$

a) Name the piece of apparatus used to transfer the sodium sulfite solution into the flask accurately.
b) Name the piece of equipment used to determine the volume of iodine required.
c) Calculate the concentration of the iodine solution.

2 Vitamin C, $C_6H_8O_6$, reacts with iodine according to the following redox equation:

$$C_6H_8O_6 + I_2 \rightarrow C_6H_6O_6 + 2H^+$$

In an experiment to determine the mass of vitamin C in a fruit juice, the following procedure was used.

$80\,cm^3$ of fruit juice was measured accurately and transferred to a $200\,cm^3$ standard flask. The standard flask was made up to the mark with water. $20\,cm^3$ portions of this solution were added to a flask and titrated with iodine, using starch as the indicator. The results of the titration with 0.01 mol l^{-1} iodine solution are shown in Table 19.3.

Table 19.3

Experiment	Volume of iodine/cm^3
1	12.5
2	12.1
3	12.0

a) Calculate the average volume of iodine used.

b) Suggest how the 20 cm^3 portion of juice was added to the flask.

c) Calculate the number of moles of iodine reacting with the 20 cm^3 juice sample.

d) Calculate the mass of vitamin C present in the original 80 cm^3 of juice.

3 Mixtures of amino acids can be separated using paper chromatography. On a chromatogram, the retention factor, R_f, for a substance can be a useful method of identifying the substance.

$$R_f = \frac{\text{distance moved by spot}}{\text{distance moved by solvent}}$$

a) A solution containing a mixture of four amino acids was applied to a piece of chromatography paper that was then placed in solvent 1. Chromatogram 1 is shown in Figure 19.8.

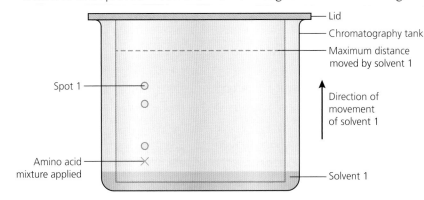

Table 19.4

Amino acid	R_f (solvent 1)
Alanine	0.51
Arginine	0.16
Threonine	0.51
Tyrosine	0.68

Figure 19.8

The retention factors for the amino acids in solvent 1 are shown in Table 19.4. Identify the amino acid that corresponds to spot 1 on the chromatogram.

b) The chromatogram was dried, rotated through 90° and then placed in solvent 2. Chromatogram 2 is shown in Figure 19.9.

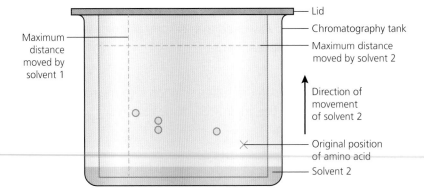

Table 19.5

Amino acid	R_f (solvent 2)
Alanine	0.21
Arginine	0.21
Threonine	0.34
Tyrosine	0.43

Figure 19.9

The retention factors for the amino acids in solvent 2 are shown in Table 19.5. Draw a circle around the spot on a copy of chromatogram 2 that corresponds to the amino acid alanine.

c) Explain why only three spots are present in chromatogram 1 while four spots are present in chromatogram 2.

Section 4 Researching Chemistry

Chapter 20

Researching chemistry

What you should know

★ Candidates must be familiar with the use of the apparatus and techniques listed in the following two tables and should be able to draw sectional diagrams for the apparatus.
★ Given a description of an experimental procedure and/or experimental results, an improvement to the experimental method can be suggested and justified.
★ Candidates must be able to process experimental results by
  ★ tabulating data using appropriate headings and units of measurement
  ★ representing data as a scatter graph with suitable scales and labels
  ★ sketching a line of best fit (straight or curved) to represent the trend observed in the data
  ★ calculating average (mean) values
  ★ identifying and eliminating rogue points
  ★ commenting on the reproducibility of results where measurements have been repeated.
★ The uncertainty associated with a measurement can be indicated in the form:
  measurement ± uncertainty.

Apparatus and techniques

As part of your Higher Chemistry experience, you should have had plenty of practice carrying out experiments using standard lab equipment and evaluating your experimental results. In the Higher exam, you are expected to be familiar with the techniques listed in Table 20.1 and the apparatus listed in Table 20.2.

Table 20.1 Some common lab techniques

Distillation
Filtration
Methods for collecting a gas: over water or using a gas syringe
Safe heating methods: using a Bunsen, water bath or heating mantle
Titration
Use of a balance, including measuring mass by difference
Determining enthalpy changes (see Chapter 17)

Table 20.2 Some items of laboratory apparatus

Beaker	Dropper	Pipette filler
Boiling tube/test tubes	Evaporating basin	Distillation flask
Burette	Funnel	Thermometer
Conical flask	Measuring cylinder	Volumetric flask
Delivery tubes	Pipette	Condenser

You should ensure you are familiar with the techniques and apparatus. The following general points are worth noting.

Pipettes and burettes

For general lab work, e.g. adding in approximate quantities of an excess acid to react with a metal, a measuring cylinder provides reasonable accuracy. The markings on a beaker are very rough but helpful for general lab work.

For analytical work, i.e. experiments carried out to determine the exact quantity or concentration of a chemical, pipettes and burettes provide much greater accuracy.

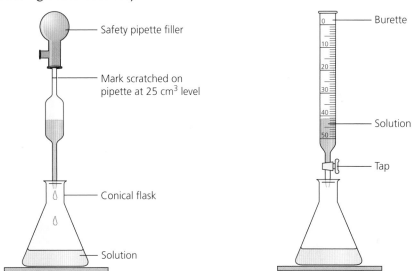

Figure 20.1 A pipette with safety filler   **Figure 20.2** A burette

A pipette is more accurate than a measuring cylinder for measuring fixed volumes of liquid. A burette can be used to measure non-standard volumes of liquid.

Example

Suggest the best piece of apparatus to measure

a) **20 cm³ of solution**
b) **23.5 cm³ of solution.**

Solution

a) pipette
b) burette

Standard flasks and standard solutions

A standard flask is used to make up a standard solution – a solution of known concentration. This is done by dissolving a known mass of solute in water and transferring to the standard flask with rinsings. Finally, the standard flask is made up to the mark with water.

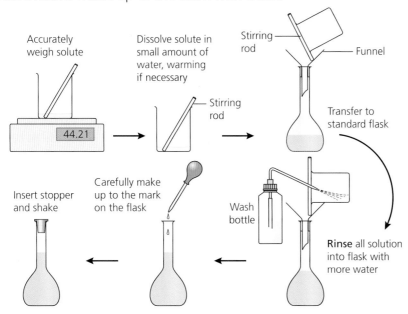

Figure 20.3 Preparing a standard solution

Collecting a gas

The two common methods for collecting gases are shown in Figures 20.4 and 20.5. The gas syringe can be used for collecting both soluble and insoluble gases and has the advantage that it can measure the volume of gas produced. Collecting a gas by bubbling through water is only appropriate if the gas is *insoluble* in water and can only be used to measure the volume if the collecting vessel is graduated, such as an upturned measuring cylinder.

Another advantage of using the gas syringe is that the gas collected will be dry, whereas it will be wet if using the collection over water method.

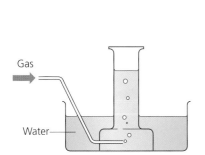

Figure 20.4 Collecting an **insoluble** gas

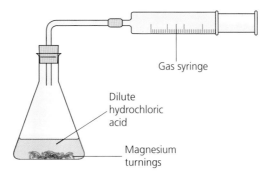

Figure 20.5 Collecting and measuring a gas using a gas syringe

Distillation

A typical distillation set-up is shown in Figure 20.6.

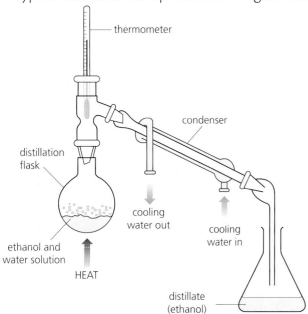

Figure 20.6 A typical distillation apparatus

Distillation is used to separate a liquid from other substances. The mixture is heated in the distillation flask (usually a round-bottomed flask to spread the heating over a larger surface area) until the boiling point of the lowest boiling liquid is reached. The vapour is then cooled as it passes through the condenser where it turns back into a liquid, which can be collected.

Methods of heating

Bunsen burners can provide rapid heating and are great for bringing solutions to boil. However, the open flame is a fire hazard (they cannot be used safely to heat flammable liquids such as alcohols) and they heat too rapidly. More accurate control can be obtained by using an electric heating mantle (which can heat to high temperatures) or a water bath. Water baths cannot provide heat beyond 100 °C. Both water baths and heating mantles (Figure 20.7) can be used for flammable liquids.

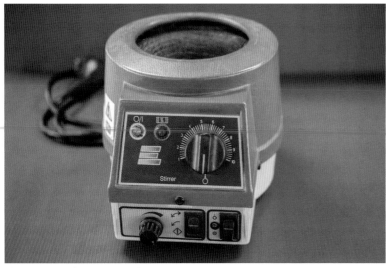

Figure 20.7 A heating mantle

Filtration

Simple filtration can be used to separate a solid from a liquid as shown in the diagram below.

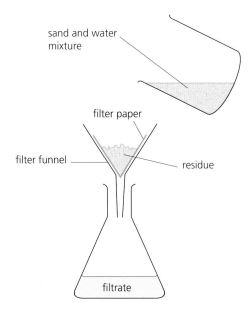

Figure 20.8 Filtration

Care must be taken not to overfill the filter funnel as this can lead to the mixture not going through the filter paper, resulting in the mixture going into the flask unfiltered.

Use of a balance

Measuring mass by difference is a common method used in laboratories and often involves the use of a weighing bottle or weighing boat. An important part of the technique is to ensure chemicals are not added to the weighing vessel while on the balance, as any spillage will contaminate the balance or could affect the accuracy of the mass.

For example, the steps below would be used to measure approximately 2 g of salt into a beaker.

1 The weighing bottle is added to the balance and weighed.
2 Approximately 2 g of salt is added to the weighing bottle.
3 The bottle is reweighed to check the mass added. If more is required, the bottle is removed from the balance and more salt added and the bottle reweighed.
4 The salt is poured from the bottle into the beaker.
5 The 'empty' weighing bottle is reweighed. By calculating the difference between the mass recorded in step 3 with the mass of the bottle (step 5) an accurate measure of the mass added will be known.

Experimental data

From your experience working with experimental data, you should know how to calculate averages, how to eliminate **rogue data**, how to draw graphs (scatter and best fit line/curve) and how to interpret graphs.

It is common in Higher exams to be presented with titration data such as that shown in Table 20.3.

Table 20.3

Titration	Volume of solution/cm³
1	26.0
2	24.1
3	39.0
4	24.2
5	24.8

Result 1 is a rough titration which is not accurate.

Results 2 and 4 could be used to calculate an average volume (= 24.15 cm³).

Result 3 is a rogue result and should be ignored.

Result 5 cannot be used to calculate the average volume as it is too far from 24.1 and 24.2 – it is not accurate.

Reproducibility

An experiment would be classed as being **reproducible** if the data obtained is the same, within experimental error, when repeated. It does not have to be the 'correct' answer. For example, if a student conducted an experiment to determine the specific heat capacity of water (which we know to be 4.18) using two methods, we could comment on the reproducibility.

Table 20.4

	Method 1	Method 2
1	6.18	4.18
2	6.19	4.98
3	6.20	4.01

We would conclude that method 1 is reproducible. Method 2 is not.

Uncertainty

Uncertainty measurements are often reported allowing you to assess the accuracy of data or the accuracy of equipment. For example, if you were told that a temperature measurement in a laboratory was 20 °C ± 10 °C you should conclude that this is highly inaccurate as it suggests an error of 10 °C, i.e. the actual temperature could be in the range 10 °C to 30 °C.

Remember

The lower the uncertainty measurement, the more accurate the data/equipment.

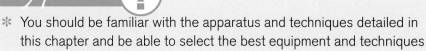

Key points

* You should be familiar with the apparatus and techniques detailed in this chapter and be able to select the best equipment and techniques for an experiment.

* You should be able to assess experimental data to understand errors, rogue data, averages and reproducibility.

Study questions

1   The following instruction was stated in a lab report:
    'Wash the zinc granules with approximately 50 cm^3 of water. Discard the washings.'
    The most appropriate piece of equipment for measuring the water is a

    A   50 cm^3 pipette   C   100 cm^3 measuring cylinder

    B   50 cm^3 burette   D   10 cm^3 syringe.

2   Lead iodide can be produced by the following reaction:
    $2KI(aq) + Pb(NO_3)_2(aq) \rightarrow PbI_2(s) + 2KNO_3(aq)$
    The most suitable method for obtaining the lead iodide is

    A   filtration   C   distillation

    B   evaporation   D   heating to dryness.

3   Two students, A and B, carried out titrations to calculate the concentration of an acid.

    Table 20.5

Student A results, cm^3	Student B results, cm^3
28.0	22.0
25.2	18.1
25.0	18.2
24.7	18.3

    a) Which student's results are the most reproducible?

    b) Calculate the average titre for each set of results.

    c) Both students used pipettes to measure the correct quantity of acid into a conical flask for the titration. Student B's pipette was labelled 20 ± 0.5 cm^3.
    Students A's pipette was found to be a more accurate pipette. Suggest a value for the uncertainty for Student A's pipette.

    d) The alkali solution used in the experiment had a concentration of 0.1 mol l^{-1} and was prepared from a stock solution which had a concentration of 1.0 mol l^{-1}.
    Describe how a pipette and standard volumetric flask could be used to accurately prepare 100 cm^3 of 0.1 mol l^{-1} alkali solution from the stock solution.

*4 A student analysed a local water supply to determine fluoride and nitrite ion levels.

    a) The concentration of fluoride ions in water was determined by adding a red coloured compound that absorbs light to the water samples. The fluoride ions reacted with the red compound to produce a colourless compound. Higher concentrations of fluoride ions produce less coloured solutions which absorb less light. The student initially prepared a standard solution of sodium fluoride with fluoride ion concentration of 100 mg l^{-1}.

    i)   State what is meant by the term standard solution.

    ii) Describe how the standard solution is prepared from a weighed sample of sodium fluoride.

    iii) Explain why the student should use distilled or deionised water rather than tap water when preparing the standard solution.

b) The student prepared a series of standard solutions of fluoride ions and reacted each with a sample of the red compound. The light absorbance of each solution was measured and the results graphed. Determine the concentration of fluoride ions in a solution with absorbance 0.012.

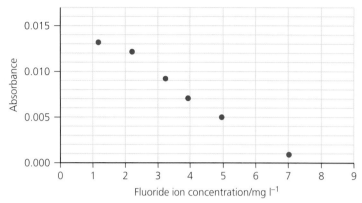

Figure 20.9

c) The concentration of nitrite ions in the water supply was determined by titrating water samples with acidified permanganate solutions. An average of 21.6 cm³ of 0.015 mol l⁻¹ acidified permanganate solution was required to react completely with the nitrite ions in a 25.0 cm³ sample of water. The equation for the reaction taking place is:

$$2MnO_4^-(aq) + 5NO_2^-(aq) + 6H^+(aq) \rightarrow 2Mn^{2+}(aq) + 5NO_3^-(aq) + 3H_2O(l)$$

Calculate the nitrite ion concentration, in mol l⁻¹, in the water. Show your working clearly.

*5 In a combustion chamber, cyanogen gas burns to form a mixture of carbon dioxide and nitrogen.

$$C_2N_2(g) + 2O_2(g) \rightarrow 2CO_2(g) + N_2(g)$$

Carbon dioxide can be removed by passing the gas mixture through sodium hydroxide solution. Copy and complete the diagram to show how carbon dioxide can be removed from the products and the volume of nitrogen gas measured.

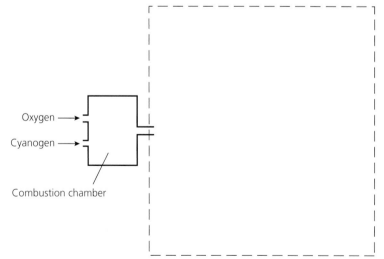

Figure 20.10

Additional features of the Higher Chemistry exam

Numeracy

The Higher Chemistry exam will contain several questions that test your numeracy skills. Some of these questions have been covered in earlier chapters, such as calculating reaction rates, enthalpy changes and percentage yield. Other questions will ask you to 'scale up' or 'scale down' as this is a skill that is used by practising scientists in their day to day job.

Being able to deal with proportion is key to answering numeracy questions in chemistry. A common layout is shown in the examples below. In all cases, the unknown (what you are being asked to calculate) should be put on the right-hand side.

> ## Example
>
> **1.2 g of methane burned to produce 52 kJ of energy. Calculate the enthalpy of combustion of methane.**
>
> ### Solution
>
> This is really a proportion question. You have to understand that the enthalpy of combustion is the energy released when 1 mole of a substance is burned completely and that 1 mole of methane is the gram formula mass, 16 g.
>
> Step 1: State a relationship.
>
> $$\text{mass} \rightarrow \text{energy}$$
> $$1.2\,\text{g} \rightarrow 52\,\text{kJ}$$
>
> (Note that energy is placed on the right-hand side as we want to calculate the energy.)
>
> Step 2: Scale to 1.
>
> $$1\,\text{g} \rightarrow \frac{52}{1.2} = 43.3\,\text{kJ}$$
>
> Step 3: Calculate for the mass you are asked for in the question.
>
> $$16\,\text{g} \rightarrow 16 \times 43.3 = 693.3$$
>
> Finally, answer the question, giving the correct units as necessary:
> The ΔH combustion for methane = **−693.3 kJ mol⁻¹**.

Example

A 10 kg batch of NaOH(s) cost £43. Calculate the cost for 4 g of NaOH(s).

Solution

Following the steps outlined in the first example:

mass → cost

10 000 g → £43

$$1\,g \to \frac{43}{10\,000} = 4.3 \times 10^{-3}$$

$$4\,g \to 4 \times (4.3 \times 10^{-3}) = £0.017 = 1.7p$$

Therefore, 4 g of NaOH(s) would cost **1.7p**.

Example

A 100 ml bottle of children's paracetamol costs £3.85. The ingredients label states that each 5 ml dose contains 120 mg of paracetamol. Calculate the cost per mg of paracetamol.

Solution

volume → mass

5 ml → 120 mg

1 ml → 24 mg

100 ml → 2400 mg

i.e. 1 bottle contains 2400 mg of paracetamol

mass → cost

2400 mg → £3.85

1 mg → £0.0016

Paracetamol costs **£0.0016 per mg**.

Example

*Theobromine, a compound present in chocolate, can cause illness in dogs and cats. To decide if treatment is necessary, vets must calculate the mass of theobromine consumed.

1.0 g of chocolate contains 1.4 mg of theobromine.

Calculate the mass, in mg, of theobromine in a 17 g biscuit of which 28% is chocolate.

Solution

A 17 g biscuit contains 28% chocolate, i.e. mass of chocolate = 28% of 17 = 0.28 × 17 = **4.76 g**

⇨
mass of chocolate → mass of theobromine

$$1.0 \text{ g} \rightarrow 1.4 \text{ mg}$$
$$4.76 \text{ g} \rightarrow 4.76 \times 1.4 = 6.66 \text{ mg}$$

There is **6.66 mg** of theobromine in the biscuit.

Example

*The maximum safe dose of lidocaine for an adult is 4.5 mg of lidocaine per kg of body mass.

1.0 cm³ of lidocaine solution contains 10 mg of lidocaine.

Calculate the maximum volume of lidocaine solution that could be given to a 70 kg adult.

Solution

mass of adult → mass of lidocaine

$$1 \text{ kg} \rightarrow 4.5 \text{ mg}$$
$$70 \text{ kg} \rightarrow 315 \text{ mg}$$

i.e. the maximum mass of lidocaine that can be given is 315 mg

mass of lidocaine → volume of solution

$$1 \text{ mg} \rightarrow 0.1 \text{ cm}^3$$
$$10 \text{ mg} \rightarrow 1.0 \text{ cm}^3$$
$$315 \text{ mg} \rightarrow 31.5 \text{ cm}^3$$

A maximum volume of **31.5 cm³** can be given to the adult.

Open-ended questions

Real-life chemistry problems rarely have a fixed answer. In the Higher exam, you will encounter questions that are open-ended; there is more than one 'correct' answer. To tackle these, look at the example shown and the marking scheme.

Example

Cooking involves many chemical reactions. Proteins, fats, oils and esters are some examples of compounds found in food. A chemist suggested that cooking food could change compounds from being fat-soluble to water-soluble.

Use your knowledge of chemistry to comment on the accuracy of this statement.

Marking scheme

0 marks: The student has demonstrated no understanding of the chemistry involved. There is no evidence that the student has recognised the area of chemistry involved or has given any statement of a relevant chemistry principle. This mark would also be given when the student merely restates the chemistry given in the question. ⇨

1 mark: The student has demonstrated a limited understanding of the chemistry involved. The candidate has made some statement(s) which is/are relevant to the situation, showing that at least a little of the chemistry within the problem is understood.

2 marks: The student has demonstrated a reasonable understanding of the chemistry involved. The student makes some statement(s) which is/are relevant to the situation, showing that the problem is understood.

3 marks: The maximum available mark would be awarded to a student who has demonstrated a good understanding of the chemistry involved. The student shows a good comprehension of the chemistry of the situation and has provided a logically correct answer to the question posed. This type of response might include a statement of the principles involved, a relationship or an equation, and the application of these to respond to the problem. This does not mean the answer has to be what might be termed an 'excellent' answer or a 'complete' one.

Author's suggested solution

To tackle a question like this, focus on the key chemical words and think about the chemistry you know. What chemical reactions do you know that involve proteins, fats, oils and esters? Can you relate this to solubility?

Proteins: Long-chain molecules linked by hydrogen bonding. Perhaps the proteins in food are insoluble as the chains are attracted to themselves. Cooking could cause the protein chains to untwist (breaking the hydrogen bonds) making them more likely to attract water to the exposed peptide links. In addition, cooking could cause the protein to hydrolyse to produce amino acids. Amino acids contain the polar amino group ($-NH_2$) and carboxyl group ($-COOH$); both can form hydrogen bonds to water, therefore the amino acids can dissolve in water.

Fats and oils: Insoluble in water as they are mainly large hydrocarbon structures. Fats and oils can hydrolyse to produce glycerol and fatty acids. Glycerol has three $-OH$ groups so it could H-bond to water molecules and dissolve. Fatty acids contain a polar head (the carboxyl group, $-COOH$) which is water soluble.

Esters: Non-polar and insoluble. Heating could hydrolyse the ester group producing an alcohol and carboxylic acid. Both of these molecules are polar and would dissolve in water.

A good answer for this question would not have to contain all of the above. Indeed, it could focus on one molecule but give lots of detail. It is also a good idea to illustrate your answer with diagrams – you could show typical structures and explain how they can bond to water. If it enhances your answer by showing the examiner that you understand the chemistry, include it!

Example ⚑

The flavour and texture of chocolate comes from a blend of compounds. Using your knowledge of chemistry, describe how you could show that there are ionic compounds and covalent compounds present in chocolate.

Solution

The following is a list of possible answers:

- Carry out an experiment to check the conductivity of molten chocolate. If ionic compounds are present, the molten chocolate should conduct electricity. Go on to discuss why ionic compounds conduct and covalent compounds do not.
- Discuss the compounds likely to be found in chocolate, for example, flavourings such as esters. Show typical ester structures and explain that they are covalent with low mp.
- Sugars are carbohydrates which are covalent.
- Aldehydes and ketones could be present; discuss this idea and draw structures.
- Antioxidants could be present to prevent spoiling. Discuss this idea and draw the structure of a typical antioxidant.
- Check the solubility of the chocolate in water and in other solvents. Describe an experiment to determine the solubility. Polar covalent and ionic compounds would dissolve in water; non-polar compounds would dissolve in non-polar solvents such as hexane. Explain why.
- Discuss electrolysis and how ionic compounds would behave.

Appendix 2

1 Structure and bonding in the first 20 elements

*1

Type		Description
1	e	monatomic gases
2	c	molecular solids
3	b	covalent network solids
4	a	metallic solids
5	d	covalent molecular gases

2 C

3 C

4 C

5 C

6 a) Monatomic gas: He, Ne or Ar; Covalent network solid: B, C or Si; Discrete covalent molecular gas: N, O, F or Cl; Discrete covalent molecular solid: S or P

 b) They have delocalised electrons.

7 Both carbon diamond and carbon graphite exist as covalent network structures.

In carbon diamond, each carbon atom is bonded to four neighbouring carbon atoms by covalent bonds. All four bonding electrons are used up forming these bonds meaning that there are no free electrons available for electrical conductivity.

In carbon graphite, each carbon atom is bonded to three neighbouring carbon atoms by covalent bonds. The remaining electron is unbonded and is contributed to a delocalised pool of electrons. These free electrons circulate between the layers of carbon atoms in graphite allowing electricity to be conducted.

2 Trends in the Periodic Table

1 A

2 B

3 C

4 B

5 a) The nuclear charge increases as you cross a period. This means that the outer electrons are more strongly attracted to the nucleus requiring more energy to remove them.

b) $Na(g) \rightarrow Na^+(g) + e^-$

c) the halogens (group 7)

6 a) P^{3-} has an extra occupied energy level compared to the aluminium ion.

b) The calcium ion has a higher nuclear charge compared to the phosphorus ion, so electrons are pulled closer to the nucleus, making the calcium ion smaller than the phosphorus ion.

3 Bonding

1 C

2 D

3 D

4 B

5 A

6 C

7 As the molecules increase in size, the boiling points increase. This is because as the molecules increase in size they have more electrons, so the strength of LDF between molecules increases. Therefore, it takes more energy to overcome the LDF and so the bp increases.

8 a) These compounds have hydrogen bonding between their molecules which is much stronger than the LDF or pdp–pdp interactions that occur between the other molecules in the graph. Consequently, these molecules have much higher melting points as more energy is required to overcome the stronger H bonds between molecules.

b) i) PH_3 – Both atoms have the same electronegativity value so this is an example of a non-polar molecule. The bonding between molecules will be London dispersion forces.

ii) H_2S – There is a difference in electronegativity values. Pdp–pdp interactions will occur between molecules.

9 Water molecules are attracted to each other by hydrogen bonding. Hydrogen sulfide molecules are attracted to each other by pdp–pdp interactions.

Since hydrogen bonding is stronger than pdp–pdp, the energy required to break hydrogen bonds is greater than the energy required to break pdp–pdp interactions. Hence, water has a higher boiling point.

*10 a) This statement is wrong as it suggests that covalent bonds are broken when covalent molecular compounds melt/boil. (This only occurs for covalent network compounds which have very high melting and boiling points, causing them to be solid at room temperature.) It is, in fact, intermolecular forces that have to be broken when covalent compounds are melted or boiled. As these forces are usually weaker than ionic bonds, covalent molecules usually have lower melting and boiling points.

b) Ionic formula refers to the ratio of ions in the lattice. In other words, there is one magnesium ion for every two chloride ions. In the lattice, the magnesium ion is likely to be surrounded by more than two chloride ions. Overall, the ratio will be 1:2.

4 Oxidising and reducing agents

1 a) $Mg \rightarrow Mg^{2+} + 2e^-$ (Oxidation); $Ag^+ + e^- \rightarrow Ag$ (Reduction)
   b) $Mg + 2Ag^+ \rightarrow Mg^{2+} + 2Ag$
   c) Mg is the reducing agent; Ag^+ is the oxidising agent.

2 fluorine, chlorine, zinc and sodium

3 a) $SO_3^{2-} + H_2O \rightarrow SO_4^{2-} + 2H^+ + 2e^-$
   b) $H_2O_2 \rightarrow O_2 + 2H^+ + 2e^-$
   c) $NO_3^- + 4H^+ + 3e^- \rightarrow NO + 2H_2O$
   d) $VO_3^- + 6H^+ + 3e^- \rightarrow V^{2+} + 3H_2O$

4 B

5 B

6 $CH_3CH_2OH + H_2O \rightarrow CH_3COOH + 4H^+ + 4e^-$

7 A

8 a) $Fe_2O_3 + 3CO \rightarrow 2Fe + 3CO_2$
   b) $Fe^{3+} + 3e^- \rightarrow Fe$
   c) reducing agent

5 Systematic carbon chemistry

1 a) A
   b) D
   c) methyl propane
   d) but-1-ene and hydrogen chloride (HCl)

2 a)

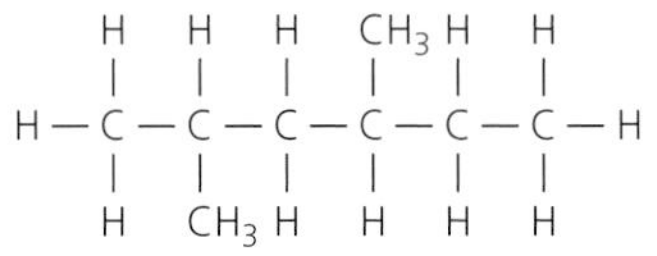

   b)

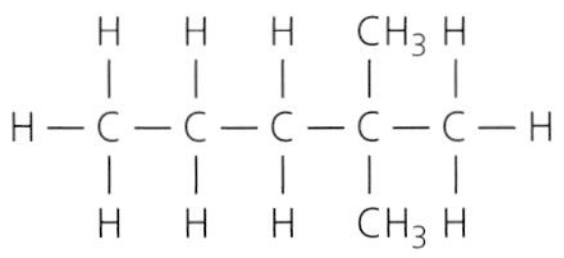

   c)

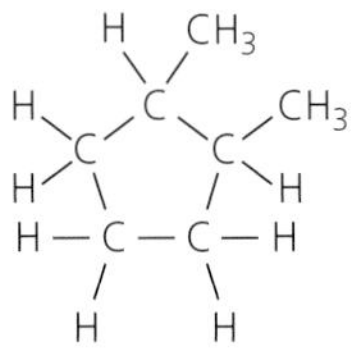

   d)

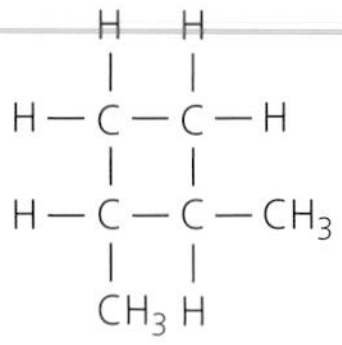

e) 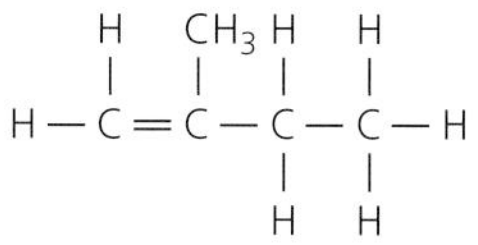
```
        H   CH₃ H   H
        |   |   |   |
   H — C = C — C — C — H
            |   |
            H   H
```

f) 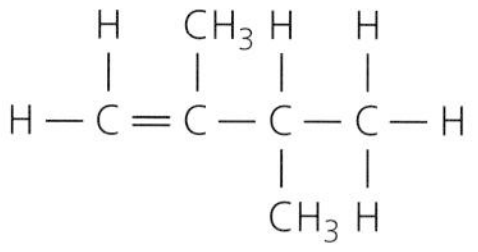
```
        H   CH₃ H   H
        |   |   |   |
   H — C = C — C — C — H
            |   |
            CH₃ H
```

g) 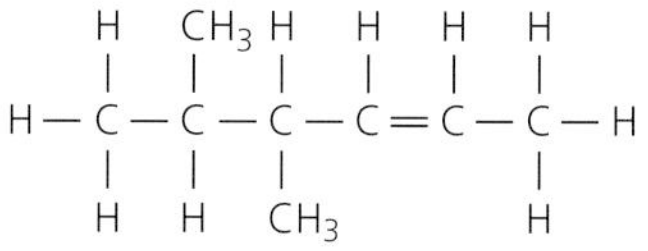
```
       H   CH₃ H   H   H   H
       |   |   |   |   |   |
  H — C — C — C — C = C — C — H
       |   |   |           |
       H   H   CH₃         H
```

h)
```
       H   CH₃      H   H
       |   |        |   |
  H — C — C = C — C = C — H
       |        |
       H        H
```

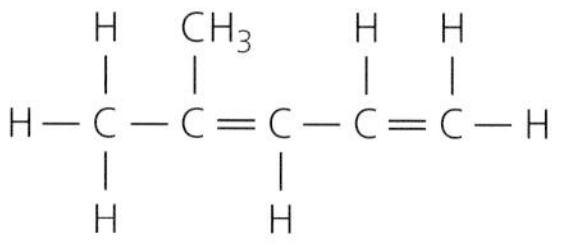

3 Draw full structural formula for compounds A−E in the following reactions:

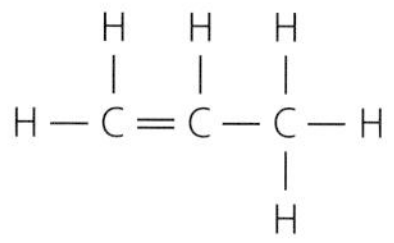

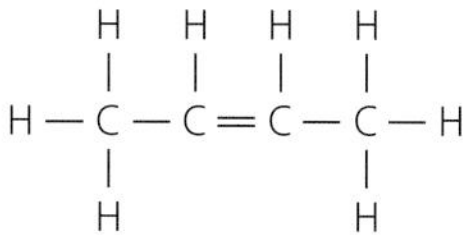

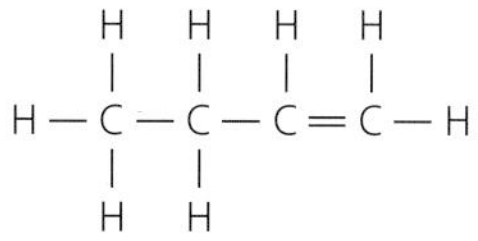

a)
```
             H   H   H
             |   |   |
   A = H — C = C — C — H
                     |
                     H
```

b)
```
             H   H   H   H                 H   H   H   H
             |   |   |   |                 |   |   |   |
   B = H — C — C = C — C — H    or    H — C — C — C = C — H
             |           |                 |   |
             H           H                 H   H
```

c)
```
            H   Cl  H
            |   |   |
   C = Cl — C — C — C — H
            |   |   |
            H   H   H
```

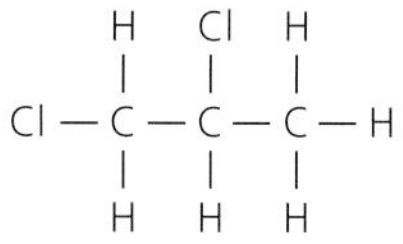

d)
```
            H   H   OH  H
            |   |   |   |
   D = H — C — C — C — C — H
            |   |   |   |
            H   H   H   H
```

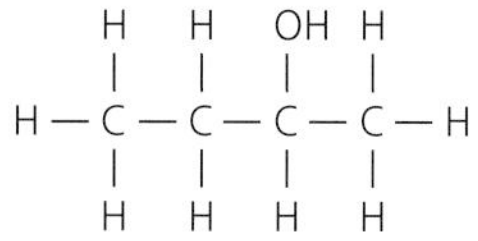

e)
```
            H   H   H
            |   |   |
   E = H — C — C — C — H
            |   |   |
            H   H   H
```

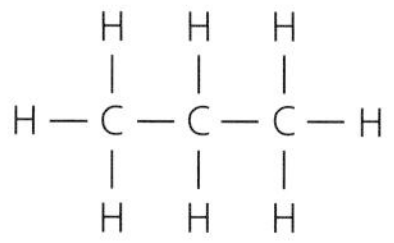

4 but-1-ene or but-2-ene

5 ethene

6 Hexane is a non-polar compound so cannot form hydrogen bonds to water molecules. It will dissolve in cyclohexane since cyclohexane is also non-polar. Cyclohexane will bond to hexane using LDF.

7 a) React the food with bromine solution. As molecules of limonene are unsaturated, we could expect a food containing limonene to rapidly decolourise bromine solution.

b) A hydrocarbon such as hexane or cyclohexane could be used. Like limonene, these solvents are non-polar so would attract limonene (using LDF) thus extracting the limonene.

c) Limonene is a bigger molecule than butane so contains more electrons. So, the strength of LDF between limonene molecules will be greater than the strength of LDF between butane molecules. It will, therefore, take more energy to break apart the LDF between limonene molecules than butane molecules so limonene will have a higher boiling point.

6 Alcohols

1 a) propan-1-ol
  b) primary
  c) 4-methylpentan-3-ol
  d) secondary
  e) 3-methylhexan-2,5-diol
  f) secondary
  g) 2-methylbutan-2-ol
  h) tertiary
  i) 4,4-dimethylpentan-1-ol
  j) primary
  k) 4-methylhexan-2,2,4-triol
  l) tertiary

2 a)
```
        H   CH3 H   H   H
        |   |   |   |   |
  HO — C — C — C — C — C — H
        |   |   |   |   |
        H   H   H   H   H
```

  b)
```
        H   CH3 H   H   H   H
        |   |   |   |   |   |
  HO — C — C — C — C — C — C — H
        |   |   |   |   |   |
        H   CH3 H   H   H   H
```

  c)
```
       H   H   H   H
       |   |   |   |
  H — C — C — C — C — H
       |   |   |   |
       OH  OH  OH  H
```

  d)
```
       H   OH  H   H   H   H
       |   |   |   |   |   |
  H — C — C — C — C — C — C — H
       |   |   |   |   |   |
       H   OH  H   H   H   H
```

  e)
```
       H   CH3 H   H
       |   |   |   |
  H — C — C — C — C — H
       |   |   |   |
       OH  OH  H   H
```

*3 a) 3,7-dimethylocta-1,6-dien-3-ol
  b) It contains an —OH which is attached to a C atom which does not have any H atoms attached.

4
```
       H   H                          H   H
       |   |    δ−    δ+   δ+   δ−     |   |
  H — C — C — O — H        H — O — C — C — H
       |   |            \        δ−   |   |
       H   H             \      /     H   H
                          \    O
                       δ+ /     \ δ+
                          H       H
```

5 Methanol molecules bond to each other by hydrogen bonding. Ethane molecules bond to each other by LDF. Since hydrogen bonding is stronger than LDF, it takes more energy to break apart the hydrogen bonding between methanol molecules than to break apart the LDF between ethane molecules. Hence, methanol has a higher boiling point.

6 The student would discover that propane-1,2,3-triol has a higher viscosity than propan-1-ol. This results from the fact that propane-1,2,3-triol has 3-OH groups per molecule whereas propan-1-ol has 1 −OH group per molecule. Consequently, propane-1,2,3-triol can form more hydrogen bonds between molecules than propan-1-ol.

7 Carboxylic acids

1 a)

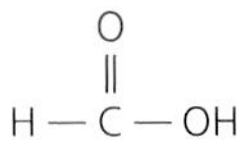

b)

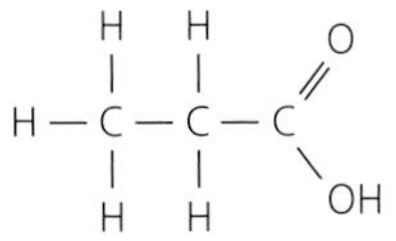

c)

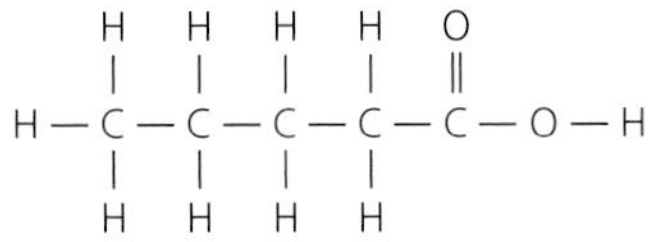

d)

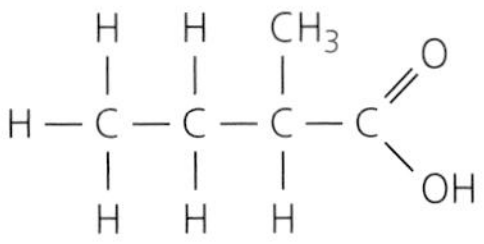

e)

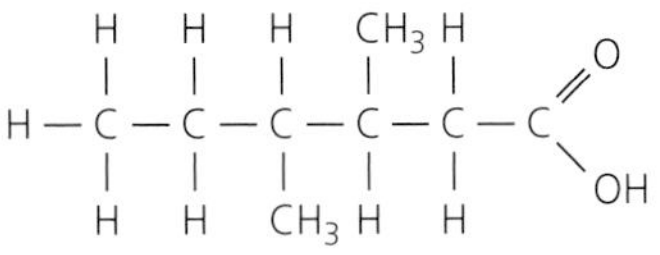

f)

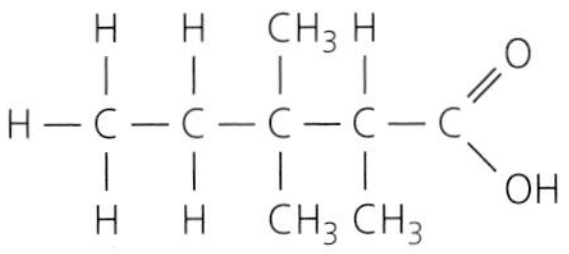

2 a) 3-methylpentanoic acid
   b) 2,3-dimethylbutanoic acid
   c) 3,3,5-trimethylhexanoic acid
   d) 3-methylpentanoic acid

3 a) magnesium butanoate + water
   b) lithium ethanoate + water
   c) sodium butanoate + water
   d) magnesium methanoate + water + carbon dioxide

4 a) lithium methanoate
   b) sodium ethanoate
   c) calcium ethanoate
   d) magnesium propanoate

5 As can be seen from the diagram, hydrogen bonds can form between water and the carboxyl group. This allows hexanoic acid to dissolve in water.

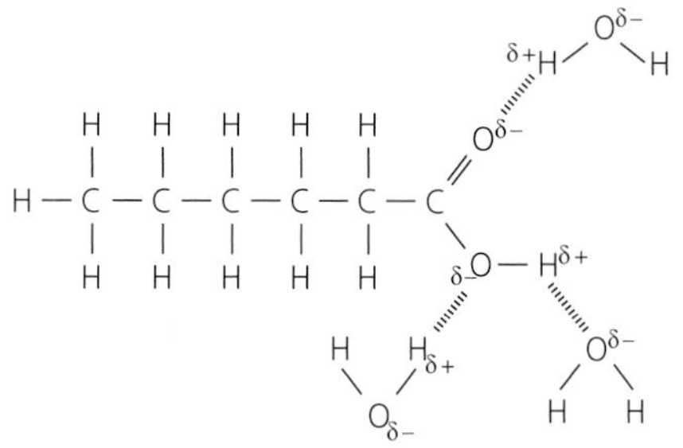

6 As can be seen from the diagrams below, ethanoic acid can form more hydrogen bonds to neighbouring ethanoic acid molecules than ethanol can form to neighbouring ethanol molecules. Consequently, it takes more energy to break the stronger hydrogen bonds between ethanoic acid molecules than is required to break the hydrogen bonds between ethanol molecules. Therefore, ethanoic acid has a higher boiling point.

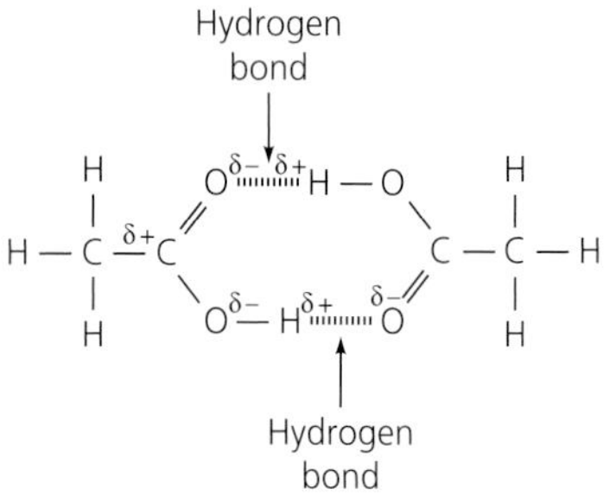

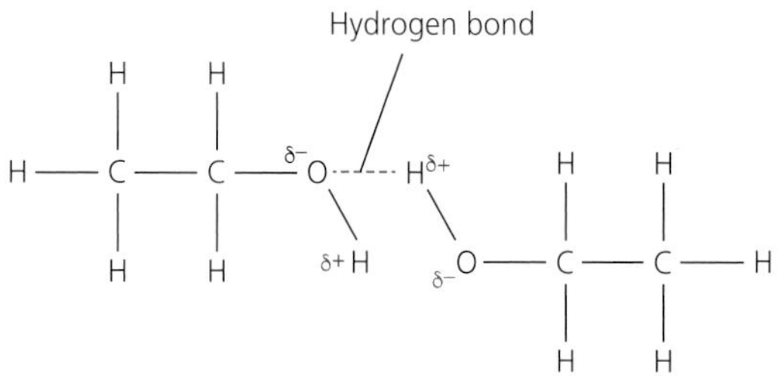

7 ammonium ethanoate

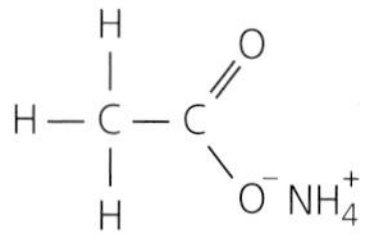

8 Esters, fats and oils

1 a)

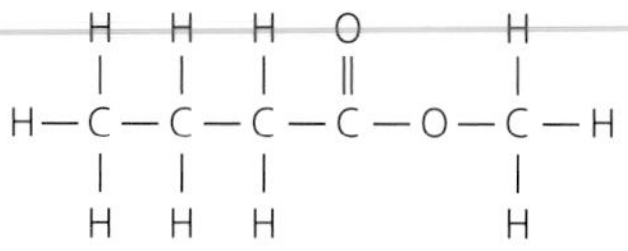

b)

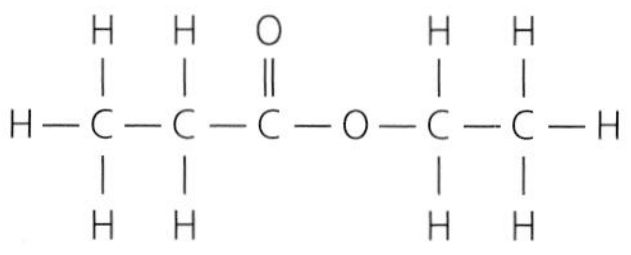

c)
```
        H   O       H
        |   ||      |
   H— C — C — O — C — H
        |           |
        H           H
```

2 a) –OH
   b)
```
        O
        ||
      — C — OH
```
   c)
```
        O
        ||
      — C — O —
```

3 a) butyl methanoate
   b) propyl pentanoate
4 a) methanol and butanoic acid
   b) ethanol and propanoic acid
   c) methanol and ethanoic acid

5

6 a) condensation
   b) hydrolysis
7 C
8 D
9 a) Vitamin C is a polar molecule as it contains lots of polar –OH
      groups. As a result, water can hydrogen bond to the –OH groups
      allowing it to dissolve vitamin C.
   b) Vitamin A is non-polar. It will not be attracted to polar water
      molecules but will be attracted to non-polar fat molecules (by
      using London dispersion forces).
10 Oils are liquids at room temperature as they have a high degree of
   unsaturation. This means that they have an irregular structure which
   does not allow the oil molecules to pack close together. Consequently,
   the molecules are not very strongly attracted to each other. Fat
   molecules have a high degree of saturation, giving them a regular
   structure which allows them to pack close together. Consequently, the
   molecules are more strongly attracted to each other and require more
   energy to pull them apart.
11 a) This molecule is a monoglyceride as it has only one ester link (OR
       only one –OH group has been used to join to a fatty/carboxylic
       acid).
    b) the long-chain hydrocarbon
12 a) addition
    b) A would have the highest degree of unsaturation. It has fewer
       hydrogen atoms than B, so the structure must be composed of
       more double bonds than B.

9 Soaps, detergents and emulsions

1 C
2 B

3 a) 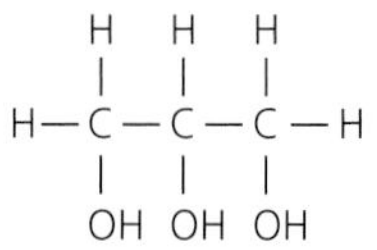
```
        H   H   H
        |   |   |
    H — C — C — C — H
        |   |   |
       OH  OH  OH
```

b) The long hydrocarbon chain is non-polar. This can attract non-polar/oily compounds. The –OH groups are polar. These can attract polar/water soluble compounds.

4 a) The circled part represents the non-polar/hydrophobic part of the detergent.

b) The non-polar chain can dissolve in the stain causing the stain to be covered in negative charges. The polar 'head' will not dissolve in the stain and, as a result, the negative charges from the polar heads will repel causing the stain to be broken up into globules of grease. These globules are now water soluble as water molecules can bond to the negative charges.

10 Proteins

1 C

2 B

*3 D

4 B

*5 C

6 C

*7 a) 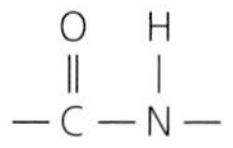
```
       O   H
       ||  |
     — C — N —
```

b) 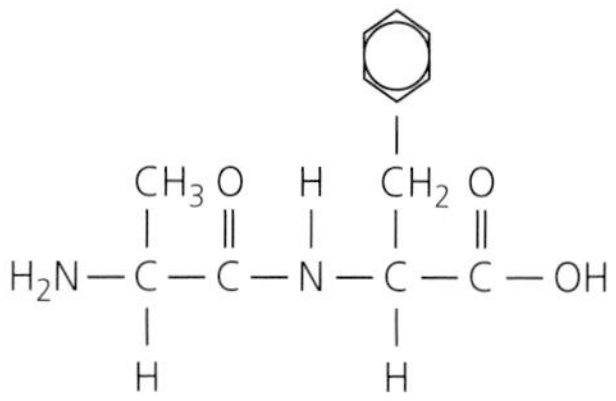
```
                        ◯
                        |
        CH₃ O   H   CH₂ O
         |   ||  |   |   ||
  H₂N — C — C — N — C — C — OH
         |           |
         H           H
```

8 a) Collagen is a protein. Heating the protein will cause it to denature.

b) amino acids

11 Oxidation of food

1 D

2 F

3 A

4 B

5 C and E

6 D

7 In addition to the –OH group which is present in both molecules, vanillin has an aldehyde group. The aldehyde group is polar ($\delta+$ C and $\delta-$ O), which will allow it to bond to polar water molecules. In eugenol, the aldehyde group is not present; it has been replaced by a hydrocarbon, which is non-polar and therefore insoluble in water.

8 a) butanone
  b) 3-methylpentan-2-one
  c) pentanal
  d) 3,3-dimethylbutanal

9 a)

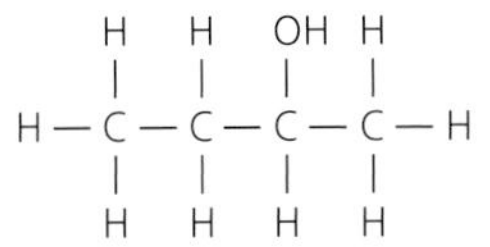

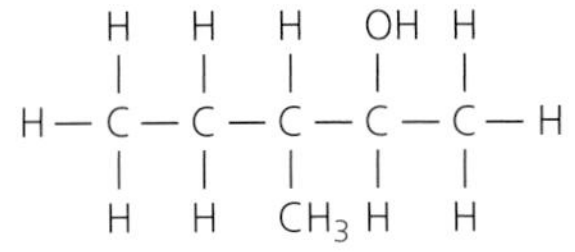

  b)

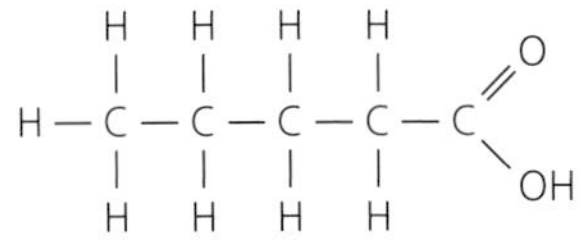

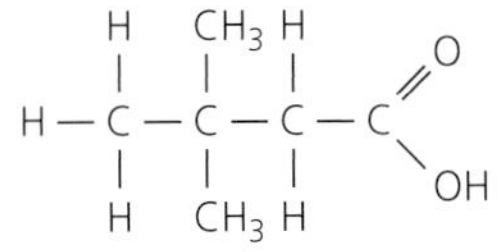

10 a) butan-1-ol

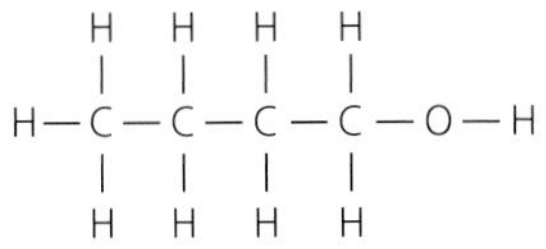

  butan-1,3-diol

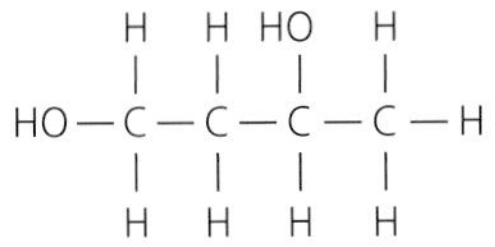

  b) butan-1,3-diol has two hydroxyl groups which allows it to form more hydrogen bonds between molecules. Consequently, more energy has to be supplied to break the hydrogen bonds between butan-1,3-diol molecules and so the boiling point is higher.

12 Fragrances

1 a) 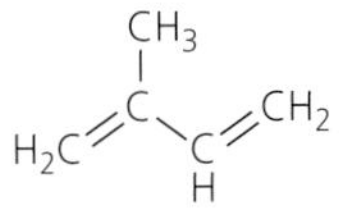
or

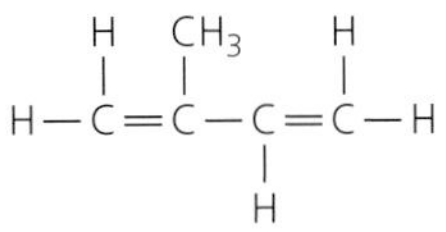

  b) 3

2 a) You could react both compounds with bromine solution. Assuming you had an equal concentration of each, squalene would require double the volume of bromine solution as it has twice as many double bonds per molecule.

  b) 6

3 a) Attempt to oxidise both with Fehling's, Tollen's or acidified dichromate. Citral would oxidise (resulting in a colour change) whereas menthone would not oxidise. (Citral is an aldehyde; menthone is a ketone.)

b)

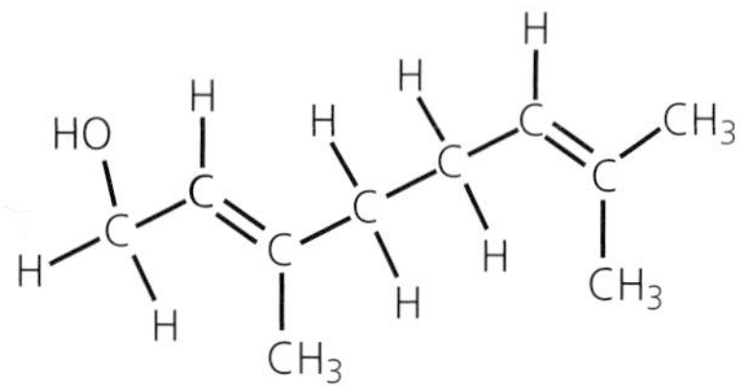

*4 C

5 A

*6 A

*7 a) Limonene is completely non-polar. Geraniol contains the polar hydroxyl group. The intermolecular bonding between geraniol molecules (H bonding) is stronger than the intermolecular bonding between limonene molecules (London dispersion forces).

b) i) aldehydes

ii)

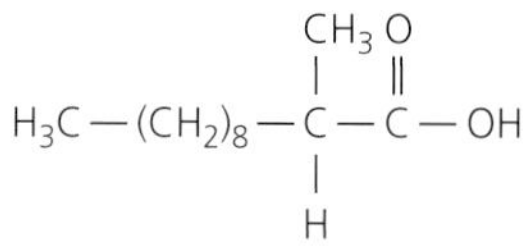

13 Skin care

1 a) A

b) B and D

c) C

*2 It can react with radicals to form stable molecules.

3 a) UV light contains enough energy to break the covalent bond between the two fluorine atoms causing the stable fluorine molecule to form two unstable fluorine free radicals.

b) Exposure to UV light can lead to skin damage (such as burning).

c) A free radical is a highly unstable atom or molecule that contains an unpaired electron.

d) Initiation: $F - F \rightarrow F^{\bullet} + F^{\bullet}$

Propagation: $C_2H_6 + F^{\bullet} \rightarrow C_2H_5^{\bullet} + H - F$

$C_2H_5^{\bullet} + F - F \rightarrow C_2H_5F + F^{\bullet}$

Termination: $F^{\bullet} + F^{\bullet} \rightarrow F - F$

$C_2H_5^{\bullet} + C_2H_5^{\bullet} \rightarrow C_4H_{10}$

$C_2H_5^{\bullet} + F^{\bullet} \rightarrow C_2H_5F$

e) Ethene contains a double bond and will undergo an addition reaction with fluorine, i.e. it is not taking part in a free radical reaction with fluorine.

14 Getting the most from reactants

1 a) Sulfur dioxide is a toxic gas. In addition, it can react with water to produce sulfuric acid which contributes to acid rain.

b) The sulfur dioxide and water could be used in step 1. The oxygen could be sold as a by-product or used in another reaction.

c) $H_2O \rightarrow H_2 + \frac{1}{2}O_2$

*2 a) 300 tonnes

b) A line from the box containing ammonia/carbon dioxide (coming from the separator) back to the reactors.

3 i) Minimise waste by designing reactions that recycle unreacted chemicals.

ii) Avoid using chemical reactions that generate toxic substances.

iii) Design products that can biodegrade.

4 a) Energy is expensive. If lots of energy is required to heat a chemical reaction, chemists may look at alternative routes that involve using a catalyst or will look to use the heat generated by exothermic reactions.

b) Many chemical reactions do not go to completion, which means that there can be unreacted chemicals left over at the end of the reaction. Rather than waste these chemicals, chemists can design processes that re-use unreacted chemicals, i.e. they are fed back into the reactor.

15 Calculations from equations

1 18 g

2 50 cm^3

3 0.91 g

4 18 g

5 6.82 g

6 a) 0.0357 mol CaO and 0.025 mol of H_2SO_4, therefore CaO is in excess as the reactant ratio is 1:1

b) 3.4 g

7 1.2 litres

8 7.77 litres

9 100 cm^3 oxygen, 300 cm^3 CO_2 and 400 cm^3 H_2O

10 8 litres

11 a) methanal

b) condensation

c) 87.4%

d) 67.8%

12 C

*13 a) It is polar/has H bonding.

b) i) methyl methanoate

ii) 58%

iii) 7.38 kg

16 Controlling the rate

1 a) $0.04\,s^{-1}$
   b) $0.25\,min^{-1}$
   c) $25\,s$
   d) $416.67\,min$

2 a) the minimum energy required for colliding particles to react OR the minimum energy required to form an activated complex
   b) an unstable, high energy, arrangement of atoms formed between reactants and products

3 a) As the diagram shows, increasing the temperature leads to all molecules having more energy. This means that there will be more molecules with energy greater than or equal to the activation energy. This is shown by the shaded area under the curve. As more molecules have the minimum energy required to react, reaction rate increases.

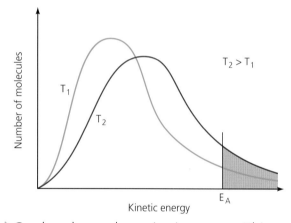

   b) Catalysts lower the activation energy. This means that many more molecules now possess the minimum energy required to react so reaction rate increases.

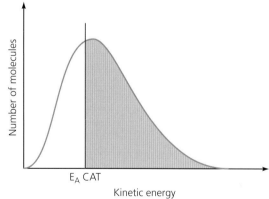

4 D
5 A
6 a) B   b) A   c) C   d) B   e) C
7 C

8 a)

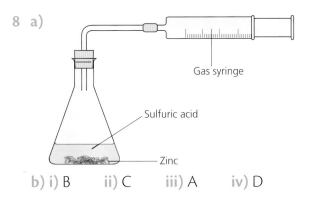

Gas syringe

Sulfuric acid

Zinc

b) i) B ii) C iii) A iv) D

17 Chemical energy

1 $-900\,kJ\,mol^{-1}$

2 $-214\,kJ\,mol^{-1}$

3 $-487\,kJ\,mol^{-1}$

4 $227\,kJ\,mol^{-1}$

5 $-124\,kJ\,mol^{-1}$

*6 $-2168\,kJ\,mol^{-1}$

7 a) $-115\,kJ\,mol^{-1}$

b) $Cl_2(g) \rightarrow Cl^\bullet + Cl^\bullet$

18 Equilibria

1 C

2 C

3 a) KOH(aq) is a source of OH^- ions. These react with H^+ ions to form water, thus removing the H^+ ions from the solution. The equilibrium will adjust to this change by shifting to the left, which results in more yellow $2CrO_4^{2-}$ being produced.

b) Sulfuric acid is a source of H^+ ions. An excess of H^+ ions causes the equilibrium to shift to the right, which results in more $Cr_2O_7^{2-}$ (orange) and H_2O being produced.

4 a) i) The reaction will speed up.

ii) The equilibrium will shift to the left causing the yield of ammonia to decrease.

b) Increasing the pressure will increase the yield of ammonia (four moles of gas on the left-hand side and two moles on the right).

c) Removing the ammonia will cause the equilibrium to shift to the right thus increasing the yield of ammonia.

5 a) PCl_3 decreases

b) PCl_3 increases

c) PCl_3 decreases

d) PCl_3 decreases

19 Chemical analysis

1 a) pipette

b) burette

c) $0.041\,mol\,l^{-1}$

2 a) $12.05\,cm^3$

b) using a pipette

c) 1.205×10^{-4} moles of iodine

d) 0.21 g

3 a) tyrosine

b)

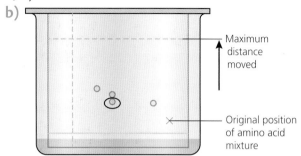

Maximum distance moved

Original position of amino acid mixture

c) With solvent 1, alanine and threonine have the same R_f value and travel the same distance as a single spot. When they are placed in solvent 2, the spot splits into two since alanine and threonine have different R_f values.

20 Researching chemistry

1 C

2 A

3 a) student B

b) student A = 25.1 cm³
student B = 18.2 cm³
Hint: When calculating average titres, only use results within 0.2 of each other. So, for student A we take the average of 25.0 and 25.2. For student B, we have three results within 0.2 of each other so we take the average of all three results.

c) You can select any value less than 0.5, e.g. ±0.4 cm³.

d) Collect a 10 cm³ pipette and transfer 10 cm³ of 1.0 mol l⁻¹ stock solution into a 100 cm³ volumetric flask. Make the flask up to the mark using deionised water.

*4 a) i) a solution of accurately known concentration

ii) The weighed sample is dissolved in a small volume of deionised water in a beaker and the solution transferred to a standard flask. The beaker is rinsed and the rinsings are also poured into the standard flask. The flask is made up to the mark, adding the last few drops of water using a dropping pipette. The flask is stoppered and inverted several times to ensure thorough mixing of the solution.

iii) Tap water contains dissolved salts that may react with sodium fluoride and affect the concentration of the solution.

b) 2 mg l⁻¹

c) concentration = 0.0324 mol l⁻¹

*5 The diagram should show the gas bubbling through NaOH (aq) to remove CO_2 (g). The nitrogen gas which bubbles through the NaOH (aq) unreacted should be measured by a gas syringe or upturned, graduated test tube/measuring cylinder in water.

Key words

Activated complex (p.118) An unstable arrangement of atoms formed at the maximum of the potential energy barrier during a reaction.

Activation energy (p.115) The energy required by colliding molecules to form an activated complex.

Addition (p.41) A reaction in which two or more molecules combine to produce a larger molecule and nothing else.

Alcohols (p.44) Carbon compounds which contain the hydroxyl functional group, –OH.

Aldehydes (p.76) Carbon compounds which contain the –CHO functional group. They are formed by oxidation of primary alcohols and they oxidise to produce carboxylic acids.

Alkanes (p.22) A **homologous series** of saturated hydrocarbons, general formula C_nH_{2n+2}. The first member is methane, CH_4.

Alkenes (p.36) A homologous series of unsaturated hydrocarbons, general formula C_nH_{2n}. Each member contains a carbon–carbon double bond. The first member is ethene, $CH_2=CH_2$.

Alkyl group A group of carbon and hydrogen atoms forming a branch in a carbon compound, for example, methyl group, CH_3–, ethyl group, C_2H_5–.

Amide link (p.71) Group of atoms formed by **condensation polymerisation** of amino acids in the formation of protein chains. The amide link is CONH and occurs between each pair of amino acid residues in the chain. Also called a peptide link.

Amino (p.71) A compound containing $-NH_2$.

Amino acids (p.71) Compounds of general formula, $H_2NCHRCOOH$ – where R is, for example, H, CH_3, $C_6H_5CH_2$ – which link by condensation reaction to form proteins. Essential amino acids cannot be synthesised by an organism and must be present in its diet.

Antioxidants (p.81) Compounds that slow oxidation reactions. They are commonly added to food to prevent edible oils becoming rancid. Examples include vitamins E and C.

Atom economy (p.109) A measure of the proportion of reactants that have been converted into products. It is calculated by using the formula, atom economy = mass of desired product/total mass of reactants $\times$ 100. Reactions with a high atom economy are desirable.

Atomic number (p.1)   The number of **protons** in the nucleus of an atom.

Biodegrade (p.94)   The breakdown of materials by bacteria or other biological means.

Bonding continuum (p.17)   A concept applied to bonding. Ionic and covalent bonding lie at opposite ends of the bonding continuum with polar covalent bonding in between.

Carbonyl group (p.78)   The carbonyl group is C=O. It is present in ketones and aldehydes.

Carboxyl group (p.50)   The functional group present in carboxylic acids, –COOH.

Carboxylic acids (p.50)   Carbon compounds which contain the carboxyl functional group. Ethanoic acid is an example of a carboxylic acid.

Catalyst (p.117)   A substance which speeds up a reaction without itself being used up. It lowers the activation energy of the reaction.

Chromatography (p.141)   A technique for separating substances. Molecules of different size or polarity can be separated by this technique which uses a **mobile phase** of gas or liquid passing over a **stationary phase** of a solid or a liquid-impregnated solid.

Closed system (p.133)   Reversible reactions will only reach a state of dynamic equilibrium when the reaction takes place in a reaction vessel which prevents reactants and products escaping.

Collision geometry (p.114)   A term used to describe the way reactants collide with each other.

Collision theory (p.114)   A theory used to explain the factors which lead to a successful reaction. It explains how altering **variables**, such as temperature, can affect the speed of the reaction. The theory requires reactants to i) collide, ii) have the correct collision geometry and iii) have a minimum energy (the activation energy) before a reaction occurs.

Concentration (p.98)   The amount of solute dissolved in a given volume of solution. The usual units are moles per litre ($mol\,l^{-1}$).

Condensation polymerisation   A process whereby many small molecules (monomers) join to form a large molecule (a **polymer**), with water or other small molecules formed at the same time. Forming a protein from amino acids is an example of condensation polymerisation.

Condensation reaction (p.56)   A reaction in which two (or more) molecules join together by the elimination of a small molecule, such as water.

Covalent bonding (p.3)   Bond formed between two atoms by the sharing of a pair of electrons. This usually occurs between non-metal atoms.

Covalent molecular (p.3)   A description of the structure and bonding in small molecules, for example, Cl_2 and H_2O.

Covalent network (p.4)   A very strong and stable structure formed by certain elements (such as B, C diamond and Si) and certain compounds (for example, SiC and SiO_2). All the atoms are held together by strong covalent bonds. Consequently, covalent network compounds are all solids at room temperature and have very high melting points.

Covalent radius (p.9)   A useful measure of atomic size, being half the distance between the nuclei of two covalently bonded atoms of an element. Covalent bond lengths between any two atoms can be obtained by adding the appropriate covalent atomic radii.

Cycloalkanes (p.36)   A homologous series of saturated ring molecules with general formula C_nH_{2n}. The simplest is cyclopropane, C_3H_6.

Dehydration   The removal of water from a single compound, for example, dehydration of ethanol, C_2H_5OH, produces ethene, C_2H_4.

Delocalised electrons (p.2)   Electrons which are not confined to a single orbital between a pair of atoms, for example, in metallic bonding. Delocalised electrons are free to move away from the atom they came from.

Denaturing/Denature (p.80)   Altering the shape of a protein by an increase in temperature or a reduction in pH. Loss of enzyme activity is one important consequence.

Detergent (p.67)   A soap-like molecule which can dissolve fats and oils. Unlike soaps, detergents do not contain a carboxylate (COO^-) head.

Displacement   A redox reaction where a metal high in the electrochemical series reacts with a metal compound lower in the electrochemical series.

Distillation (p.150)   A process used for separating liquid mixtures. A liquid is boiled and its vapour then condensed to collect pure samples of the liquid.

Electrochemical series (p.29)   A list of chemicals arranged in order of their increasing ability to gain electrons, in other words in order of increasing oxidising power.

Electrolysis (p.159)   The process which occurs when a current of electricity is passed through a molten electrolyte (resulting in decomposition) or an electrolyte solution (which results in decomposition of the solute and/or the water).

Electron (p.2)   A particle which moves around the nucleus of an atom. It has a single negative charge but its mass is negligible compared to that of a **proton** or **neutron**.

Electronegativity (p.11) The strength of the attraction by an atom of an element for its bonding electrons. If the electronegativities of two atoms sharing electrons are similar, the bond will be almost purely covalent. The greater the difference in electronegativities, the more likely the bond is to be polar covalent or even ionic.

Emulsifier (p.67) A compound which allows oil and water to mix. Emulsifiers contain a hydrophillic head and a hydrophobic tail.

Emulsion (p.67) A mixture of liquids where small droplets of one liquid are dispersed in another liquid. Emulsions of oil and water are commonly found in food.

Endothermic reaction (p.116) A reaction in which heat energy is absorbed from the surroundings. It has a positive enthalpy change (ΔH).

End-point (p.141) The point in a titration where the indicator changes colour to indicate that the reaction is complete.

Enthalpy change (p.116) The difference in heat energy between reactants and products in a reaction.

Enthalpy of combustion (p.124) The enthalpy change when one mole of a substance is completely burned in oxygen.

Enzyme (p.71) A globular protein which is able to catalyse a specific reaction.

Equilibrium (p.133) State attained in a reversible reaction when forward and reverse reactions are taking place at the same rate.

Essential amino acids (p.72) Amino acids which cannot be made by the body. They must be obtained from the diet.

Essential oils (p.84) Oils extracted from plants. They usually have distinctive smells, are non-polar, volatile and contain compounds known as terpenes.

Esters (p.56) Carbon compounds formed when alcohols react with carboxylic acids by condensation.

Exothermic reaction (p.116) A reaction in which heat energy is released to the surroundings. It has a negative enthalpy change (ΔH).

Fats (p.60) Esters formed from one molecule of glycerol and three molecules of, usually saturated, long-chain carboxylic acids. The compounds have melting points high enough to be solid at room temperature. See also **oils.**

Fatty acids (p.60) Carboxylic acids formed from the hydrolysis of fats and oils.

Feedstock (p.93) A substance derived from a **raw material** which is used to manufacture another substance.

Free radicals (p.89) Highly reactive atoms or molecules with unpaired electrons.

Free radical scavenger (p.91)   A compound added to plastics, cosmetics and foods to prevent free radical reactions. These scavengers react with free radicals to produce stable molecules. This terminates the reaction.

Functional group (p.44)   A group of atoms or type of carbon–carbon bond which provides a series of carbon compounds with its characteristic chemical properties, for example, –CHO, –C=C–.

Gas liquid chromatography (GLC) (p.140)   A technique used to separate mixtures in the gas phase.

Glycerol (p.24)   Propane-1,2,3-triol; formed from the hydrolysis of fats and oils.

Group (p.9)   A column of elements in the Periodic Table. The values of a selected physical property show a distinct trend of increase or decrease down the column. The chemical properties of the elements in the group are similar.

Haber process (p.138)   The industrial production of ammonia from nitrogen and hydrogen using high pressure and temperature, with iron as a catalyst.

Hess's law (p.126)   The enthalpy change of a chemical reaction depends only on the chemical nature and physical state of the reactants and products and is independent of any intermediate steps.

Homologous series (p.36)   A group of chemically similar compounds which can be represented by a general formula. Physical properties change progressively through the series, for example, the alkanes, general formula C_nH_{2n+2}, show a steady increase in boiling point.

Hormones (p.70)   Chemicals, often complex proteins, which regulate metabolic processes in the body. An example is insulin which regulates sugar metabolism.

Hydration   The addition of water to an unsaturated compound, for example, the hydration of ethene, C_2H_4, produces ethanol, C_2H_5OH.

Hydrocarbon (p.22)   A compound containing the elements carbon and hydrogen only.

Hydrogenation   The addition of hydrogen to an unsaturated compound; for example, hydrogenation converts alkenes to alkanes and oils into fats.

Hydrogen bonds/bonding (p.22)   Intermolecular forces of attraction. The molecules must contain highly polar bonds in which hydrogen atoms are linked to very electronegative nitrogen, oxygen or fluorine atoms. The hydrogen atoms are then left with a positive charge and are attracted to the electronegative atoms of other molecules. They are a specific, stronger type of permanent dipole–permanent dipole interaction.

Hydrolysis (p.59)   The breaking down of larger molecules into smaller molecules by reaction with water.

Hydrophilic (p.66)   A term used to describe molecules, or parts of a molecule, which are attracted to water. For example, the –OH group in alcohols is hydrophilic.

Hydrophobic (p.66)   A term used to describe molecules, or parts of a molecule, which repel water and will not bond to water. For example, the long hydrocarbon chains in fats and oils are hydrophobic.

Hydroxyl group (p.24)   The –OH group; it is found in alcohols.

Indicator (p.141)   A chemical dye added to a titration to detect the end-point.

Initiation (p.90)   The start of a free radical chain reaction that involves the breaking of a covalent bond to form free radicals, e.g. $Cl_2 \rightarrow Cl + Cl$.

Intermolecular bonds/bonding (p.4)   Bonds between molecules, such as London dispersion forces, permanent dipole–permanent dipole interactions and hydrogen bonds. They are much weaker than **intramolecular** bonds.

Intramolecular bonds (p.4)   Bonds within molecules, such as covalent and polar covalent bonds.

Ion–electron equation (p.81)   Equation which shows either the loss of electrons (oxidation) or the gain of electrons (reduction).

Ionic bond (p.17)   Bond formed as a result of attraction between positive and negative ions.

Ionisation (p.13)   The loss or gain of electrons by neutral atoms to form ions, for example,

$$Na(g) \rightarrow Na^+(g) + e^-$$
$$Cl(g) + e^- \rightarrow Cl^-(g)$$

'Ionisation enthalpy' is usually reserved for enthalpy changes referring to the formation of positive ions.

Ionisation energy (p.13)   The energy required to remove 1 mole of electrons from 1 mole of atoms in the gaseous state.

Ions (p.2)   Atoms or groups of atoms which possess a positive or negative charge due to loss or gain of electrons, for example, Na^+ and CO_3^{2-}.

Isomers (p.38)   Compounds which have the same molecular formula but different structural formulae.

Isoprene (p.84)   A five-carbon compound that forms the basis of all terpenes. It is 2-methylbuta-1,3-diene.

Isotopes   Atoms of the same element which have different numbers of neutrons. They have the same atomic number but different **mass numbers**.

Ketones (p.76)   Carbon compounds which contain the carbonyl group (C=O). They are formed from the oxidation of secondary alcohols. Unlike aldehydes, ketones cannot be oxidised using mild oxidising agents.

Lattice (p.18)   The three-dimensional arrangement of positive and negative ions in the solid, crystalline state of ionic compounds.

Le Chatelier's principle   If any change of physical or chemical conditions is imposed on any chemical equilibrium then the equilibrium alters in the direction which tends to counteract the change of conditions.

Limiting reagent (p.101)   The reactant which is not in excess. This controls the amount of product formed.

London dispersion force (p.4)   A force of attraction between all atoms and molecules formed from temporary and induced dipoles.

Mass number   The total number of protons and neutrons in the nucleus of an atom.

Metallic bonding (p.2)   The bonding responsible for typical metallic properties such as malleability, ductility and electrical conductivity in metals and alloys. Each atom loses its outer electrons to form positive ions. These ions pack together in a regular crystalline arrangement with the electrons delocalised through the structure, binding the ions together.

Miscibility (p.44)   The ability of liquids to mix perfectly together. In contrast, immiscible liquids form clearly defined layers with the denser liquid forming the lower layer.

Mobile phase (p.176)   In chromatography, the moving part of the process; for example, the inert gas in GLC which carries the mixture of compounds through the column, or the solvent in paper chromatography which carries the mixture of compounds up the paper.

Molar bond enthalpy (p.131)   The energy required to break one mole of covalent bonds. Values are listed in the data booklet.

Molar volume (p.103)   The volume of one mole of a gas at a specified temperature and pressure.

Mole (p.13)   The gram formula mass of a substance. It contains 6.02×10^{23} formula units of the substance. The commonly used abbreviation for mole is 'mol'.

Molecular formula (p.36)   Formula which shows the number of atoms of the different elements which are present in one molecule of a substance.

Molecule (p.115)   A group of atoms held together by covalent bonds.

Monatomic (p.5)   A term used to describe the noble gases since they are composed of individual atoms which do not bond to each other. They are held together by London dispersion forces in the liquid and solid state.

Neutron (p.177)   A particle found in the nucleus of an atom. It has the same mass as a proton but no charge.

Non-polar covalent bond (p.16)   A covalent bond where both atoms share the electrons equally. This occurs between all elements that exist as molecules, such as Cl_2 and S_8, since the atoms joining are identical. It also occurs in compounds where the bonding atoms have a small difference in electronegativity, such as hydrocarbons.

Nucleus (p.1)   The extremely small centre of an atom where the neutrons and protons are found.

Oils (p.60)   Esters formed from one molecule of glycerol and three molecules of, usually unsaturated, carboxylic acids. Oils have melting points low enough to be liquid at normal room temperature. See **fats**.

Oxidation (p.76)   A process in which electrons are lost.

Oxidising agent (p.29)   A substance which gains electrons, in other words is an electron acceptor.

Peptide link (p.71)   See **amide link**.

Percentage yield (p.106)   This is the actual yield of substance obtained divided by the theoretical yield calculated from the balanced equation then multiplied by 100.

Period (p.9)   A horizontal row in the Periodic Table.

Periodic Table (p.1)   An arrangement of the elements in order of increasing atomic number, with chemically similar elements occurring in the same main vertical columns (groups).

Permanent dipole–permanent dipole interactions (p.21)   The attraction between molecules which possess a permanent dipole because of the presence of polar bonds.

pH (p.121)   A measure of the acidity of a solution.

Polar covalent bond (p.17)   A bond formed between non-metallic atoms by sharing a pair of electrons. If the atoms have considerably different electronegativities, the electrons are not shared equally, the more electronegative atom becoming slightly negative in comparison to the other atom. As a result the bond is 'polar', for example, $H^{\delta+}-Cl^{\delta-}$.

Polymer (p.176)   A very large molecule which is formed by the joining together of many smaller molecules (monomers).

Polymerisation (p.175)   The process whereby a polymer is formed.

Propagation (p.90)   The second stage in a free radical chain reaction where one free radical reacts with a molecule to produce another free radical.

Proton (p.177)   A particle found in the nucleus of an atom. It has a single positive charge and the same mass as a neutron.

Rate of reaction (p.114)   A measure of the speed of a chemical reaction.

Raw material (p.94)   A useful substance for the chemical industry which found naturally, for example, crude oil, water, air, metallic ores, coal, etc. Feedstocks are obtained from raw materials.

Redox reaction (p.28)   A reaction in which reduction and oxidation take place. Electrons are lost by one substance and gained by another.

Redox titration (p.145)   An experiment in which the volumes of aqueous solutions of a reducing agent and an oxidising agent, which react together completely, are measured accurately. The concentration of one of the reactants can then be determined provided the concentration of the other reactant is known.

Reducing agent (p.29)   A substance which loses electrons, in other words an electron donor.

Reduction (p.80)   A process in which electrons are gained.

Relative rate (p.112)   Reciprocal of time, i.e. $\dfrac{1}{time}$

Reproducibility (p.152)   Results obtained from an experiment are said to be **reproducible** if the same data can be obtained when the experiment is repeated. An experiment with good reproducibility will produce the same results when carried out again and again.

Retention time (p.140)   The length of time it takes a substance to reach the detector, in a chromatography experiment, after being injected into the chromatography column.

Reversible reaction (p.133)   One which proceeds in both directions, for example:

$$N_2 + 3H_2 \rightleftharpoons 2NH_3$$

Rogue data (p.151)   Results obtained from an experiment that are unusual/do not fit the pattern of expected results. Usually caused by experimental error.

Saturated compound (p.38)   A compound in which all carbon–carbon covalent bonds are single bonds.

Screening (p.11)   The ability of electrons in the inner energy levels of an atom to reduce the attraction of the nuclear charge for the electrons of the outermost levels.

Spectator ion   An ion which is present in a reaction mixture but takes no part in the reaction.

Standard solution (p.141)   A solution of known concentration.

State symbols (p.97)   Symbols used to indicate the state of atoms, ions or molecules: (s) = solid; (l) = liquid; (g) = gas; (aq) = aqueous (dissolved in water).

Stationary phase (p.176)   In chromatography, the phase other than the mobile phase. For example, the liquid in GLC.

Structural formula (p.36)   A formula which shows the arrangement of atoms in a molecule or ion. A full structural formula shows all of the bonds. A shortened structural formula shows the sequence of groups of atoms.

Temporary dipole (p.20)   Formed in all atoms where an excess of electrons is formed at one part of the atom. Temporary dipoles are the basis for London dispersion forces.

Termination (p.90)   The final stage in a free radical chain reaction where two free radicals combine to form a stable molecule.

Terpene (p.84)   Unsaturated compounds found in many plant oils. They are formed from the joining together of isoprene units.

Transition metals   The elements which form a 'bridge' in the Periodic Table between groups II and III; for example, iron and copper.

Triglyceride (p.61)   The molecules found in fats and oils. They are formed from one glycerol molecule joining to three fatty acid molecules.

Ultraviolet light (p.89)   A high-energy form of radiation which can break bonds in molecules, causing free radicals to form.

Unsaturated compound (p.38)   A compound in which there are carbon–carbon double or triple bonds, such as alkenes, alkynes and vegetable oils.

Van der Waals' forces (p.20)   The forces of attraction that occur between all atoms and molecules. They are known as intermolecular forces and include hydrogen bonding, permanent dipole–permanent dipole interactions and London dispersion forces. Van der Waals' forces are much weaker than covalent bonds.

Variable (p.176)   Something that can be changed in a chemical reaction, such as temperature, particle size, concentration, etc.

Viscosity/viscous (p.24)   A description of how 'thick' a liquid is, for example, engine oil is 'thicker' (more viscous) than petrol.